# THE ECONOMICS OF ENERGY

# THE ECONOMICS OF ENERGY

## MICHAEL G. WEBB

and

## MARTIN J. RICKETTS

A HALSTED PRESS BOOK

JOHN WILEY & SONS
New York

© Michael G. Webb and Martin J. Ricketts 1980

First published in Great Britain 1980 by
The Macmillan Press Ltd

Published in the U.S.A. by
Halsted Press, a Division of
John Wiley & Sons, Inc.
New York

**Printed in Hong Kong**

Library of Congress Cataloging in Publication Data

Webb, Michael Gordon.
 The economics of energy.

 "A Halsted Press book."
 Includes bibliographical references and index.
 1. Energy policy. 2. Economic policy.
3. Power resources. I. Ricketts, Martin J.,
joint author. II. Title.
HD9502.A2W4   1980   333.7   79–18708
ISBN 0 470–26841–7

To Diana and Veronica

# CONTENTS

viii                                 *Contents*

# PREFACE

In recent years there has been a proliferation of books dealing with energy and particular aspects of energy economics. For the most part these books have been concerned with either particular energy industries, the general planning and formulation of energy policy or with the use and effects of specific policy instruments, such as the taxation of North Sea oil in the United Kingdom or the Federal regulation of interstate gas prices in the United States. It has been our objective in writing this book to provide an integrated and reasonably comprehensive introduction to the economics of energy. We chose not to devote separate chapters to the individual energy industries in the belief that this approach obscures the common set of problems which are associated with the development and use of the different forms of energy.

The analytical framework of the book is that of Paretian welfare economics. It is aimed primarily at undergraduates' reading economics who have studied intermediate-level microeconomics. Thus we hope that it will be especially useful for students studying public policy economics, natural resource economics, applied economics and environmental economics. By making the book comprehensive and by providing extensive references we have tried to make most of it accessible to students in other disciplines, such as engineering and geography, who are studying topics within the subject area of energy economics. Similarly most of it should be comprehensible to those government officials who are concerned with questions of energy policy.

We would like to thank the Editor of *Energy Policy* for granting permission for us to use material in Chapters 4 and 8 from two articles by Michael Webb which were originally published in that journal. In addition we would like to thank Barbara Dodds, Barbara Jacques, Gail Shepherd, Tina Asu and Linda Waterman for typing the manuscript.

*February 1979*

MICHAEL G. WEBB
MARTIN J. RICKETTS

# ACKNOWLEDGEMENTS

The authors and publishers wish to thank the following, who have kindly given permission for the use of copyright material:

Ballinger Publishing Company, for a table from *Environmental Management*, ed. G. F. Rohrlich, © 1976.

The Controller of Her Majesty's Stationery Office, for tables based on statistics from official publications.

The Ford Foundation, for a table in *A Time to Choose*, Energy Policy Project of the Ford Foundation, published by Ballinger Publishing Company, 1974, © The Ford Foundation.

Institute of Social and Economic Research, University of Alaska, for a table from *Alaska's Petroleum Leasing Policy*, by Gregg K. Erickson.

International Communication Agency, for a table from the *National Energy Plan*, Executive Office of the President, The White House, 29 April 1977.

Lexington Books, for an adaptation of a table in *Economic Aspects of the Energy Crisis*, by Harry W. Richardson (Lexington, Mass.: Lexington Books, D. C. Heath & Co., © 1975, D. C. Heath & Co.).

National Coal Board, for a table from *National Coal Board Annual Reports and Accounts*.

Shell International Petroleum Co. Ltd, for tables from 'Energy in Perspective' (1977) and 'A Generation of Change: World Energy Patterns 1950–1975' (1976), published in the *Shell Briefing Service*.

South Coast Air Quality Management District, for a table from *1974 Profile of Air Pollution Control*, published by The Los Angeles Air Pollution Control District.

Every effort has been made to trace all the copyright-holders, but if any have been inadvertently overlooked the publishers will be pleased to make the necessary arrangement at the first opportunity.

# 1. INTRODUCTION

Energy has been succinctly defined as 'the ability to do work'. It exists in a variety of forms, mechanical or kinetic, electrical, thermal, radiant, chemical or nuclear, and all changes in a physical state, for example in production, require an input of one or more of these forms of energy. These two simple statements should be sufficient to indicate the fundamental significance of energy, and the economist who is concerned with the allocation of scarce resources between competing uses will inevitably have to consider either explicitly or implicitly, as part of this allocation problem, the energy transformations involved in the processes of production and consumption.

For any study of energy it is necessary to be aware of two very important laws of physics, the first and second Laws of Thermodynamics. Simply expressed, the first states that energy can neither be created nor destroyed. This is sometimes referred to as the law of 'the conservation of energy'. The second states that, though energy may be transformed from one state to another, such transformations can never be one hundred per cent efficient. Part of the energy is inevitably lost as unavailable, and unusable, heat. The chemical energy locked up in coal, for example, may be burnt to product heat which can transform water into steam which may then be used to turn a turbine and hence to generate electricity. This electricity can then illuminate a light bulb or drive an electric train. However, during each of these transformations there is always a loss of energy as heat.

In spite of its profound importance to productive activity and indeed to the continuance of life itself, one might still inquire whether economics is of any special relevance to the study of energy. Certainly the title of this book, *The Economics of Energy*, is not intended to convey the impression that the Principles of Economics are different, or that they suddenly go into abeyance, when problems of energy supply or use are being considered. In general the interest of economists in energy does not stem from the fact that different principles are involved, but from the recognition that some of the major problems associated with energy production and use require complex analytical tools to help in yielding solutions, tools which are still being developed in the economics literature. Three particularly intractable problem areas are intimately bound up with energy economics: the problems of externality, of uncertainty and of equity. These three areas are precisely those to which economists have devoted an increasing amount of attention in recent years.

*Externality*

The external effects of energy production, transportation and use are clearly of great public interest. Air pollution from motor-car exhaust gases or from power stations and other plants involving the burning of fossil fuels, noise from motorways and airports, water pollution from the spillage of oil at sea, visual disamenity stemming from the strip mining of coal, all these are examples of the types of external costs which are frequently associated with energy use. External costs, however, do not exclusively take the form of environmental pollution. The discovery and extraction of fossil-fuel deposits, especially oil, can also give rise to important externalities. Information about geological conditions, for example, is a valuable product in itself and one which it is difficult to keep from potential competitors. Exploration activity on the part of one firm may therefore confer external benefits on other firms in the form of information.

*Uncertainty*

As with the problem of externalities the existence of uncertainty is not something which is of exclusive concern for energy economics. However, it does pose particularly acute problems in this field. Decisions about the rate of depletion of oil and gas reserves, about the appropriate prices to charge for different fuels, or the size and type of investment in new power stations, all have to be taken on the basis of incomplete information. The severity of the problem can be illustrated simply by considering a few of the major uncertainties.

Knowledge of fossil-fuel reserves, their size, type and location is difficult to estimate with any degree of accuracy. The estimates that are available are in reality little more than informed guesses. Technological change, which can have profound implications for resource depletion, investment and pricing policy as well as environmental policy, is obviously shrouded in uncertainty. What are the chances of solving the environmental problems of nuclear reactors, such as the disposal of high-level wastes or the protection of plutonium stocks from theft? What is the probability of developing fusion power within the next fifty years on a commercial scale? What are the prospects for alternative-energy sources such as solar, wind, wave, tidal or geothermal energy? What improvements can be expected in the efficiency of conventional fuel use and in pollution abatement technology, and what will be their costs? Political change can also add to the uncertainties surrounding energy supply. Will the Organization of Petroleum Exporting Countries (OPEC) maintain its cartel or will it prove unstable and eventually break up? These uncertainties relate mainly to the supply side of the energy market, but similar questions can be asked about demand. Changes in the demand for energy will depend, *inter alia*, upon income changes, technological and political developments, all of which are subject to great uncertainty.

## Equity

The problem of equity has at least three facets in connection with energy: intra-national, international and inter-generational. The first of these refers to the distribution of access to energy between individuals in a given country. We cannot devote a great deal of attention here to a discussion of why energy is considered 'special' in this respect. In market economies income is what ultimately determines the ability of consumers to demand any particular services, and thus access to energy is merely a part of this much wider distributional problem. However, whether the reasons are sound or otherwise, much of the redistribution which occurs in Western countries is in kind rather than cash. Energy is often viewed as a 'necessity' in the same way as housing and health, and is of considerable concern to designers of social security systems.

Thus much controversy surrounds decisions to disconnect electricity supplies from poor consumers who are unable to pay and schemes for 'energy rebates' and 'inverted price tariffs' have been devised and implemented in a number of countries.

Energy is certainly a 'necessity' in the technical sense that the income elasticity of demand is less than unity, so that the proportion of total income used to purchase energy falls as income rises. Cross-section data reported in Table 1.1 indicate that the percentage of total income spent on energy falls from 15.2 per cent for the poor in the United States to 4.1 per cent for the well-off. It is this aspect of the demand for energy which makes energy pricing so sensitive from a distributional point of view. To raise the price of energy is akin to taxing a commodity which appears prominently in the budgets of the poor.

TABLE 1.1

*Expenditure on Energy – United States, 1972–3*

| Income status | Average income | Average annual b.t.u. | Total annual income spent on energy |
|---|---|---|---|
| | $ | (millions) | % |
| Poor | 2,500 | 207 | 15.2 |
| Lower-middle | 8,000 | 294 | 7.2 |
| Upper-middle | 14,000 | 403 | 5.9 |
| Well-off | 24,500 | 478 | 4.1 |

SOURCE: *A Time to Choose.* Energy Policy Project of the Ford Foundation (1974) table 26, p. 118.

A similar problem arises in the international context. Rising energy prices, especially of oil, can have serious repercussions on the development prospects of many very poor states. Those with limited indigenous energy supplies therefore face a deterioration in their terms of trade and must make many of the adjustments forced on richer states, but from a much lower starting-point.

Finally, present pricing and investment decisions will affect the rate of resource depletion and the state of the natural environment. These decisions will therefore affect the type of world inherited by future generations. How far the interests of the future should be considered by present decision-makers is clearly an ethical question of great significance, and the answer to it could have a considerable impact on the use of energy resources.

The existence of the problems outlined above produces two important results. First, it leads to demands that the state should take an active interest in developing an energy policy. Second, it makes it very difficult to decide, by the very complexity and nature of the problems, what the state is capable of doing about them. Clearly it would be unrealistic to expect a study of economics to product all the answers. The range of issues is so great that many disciplines are involved. Scientists and engineers will be concerned with the technical problems of energy production, ecologists with the environmental implications, philosophers with ethical questions and sociologists and political scientists with the social and political consequences of the use of various forms of energy.

Economics, however, has a very important role to play. In particular, the use of economic models helps to trace the ramifications of various ethical or technical assumptions. It enables us to ask questions such as, what will be the effect on the rate of depletion of fossil fuel of technical advance in the field of nuclear energy? Or alternatively we might ask, what *should* be the impact on depletion rates if we accept the ethical position that the preferences of future generations are not to be discounted? If ecologists or sociologists inform the world about the particularly pernicious consequences of a form of pollution, further questions might be asked. How can the damage be assessed? By how much should the pollution be reduced? What instruments are at the Government's disposal for achieving such a reduction, and what are their additional effects, if any? Where uncertainty poses severe problems the economist has to consider what criteria exist for rational choice under these conditions and what assumptions underlie them.

The main justification for the development of analytical models by economists, it should be noted, is not that they provide a description of the world, but that they give us a way of looking at it. Economics is an 'engine of analysis', to use Keynes's phrase. In similar vein D. H. Robertson has written that 'all our analytical models, from the simplest to the most elaborate, are only aids to thought in dealing with the complexities of the real world'. Among other things, we present and consider some of the models developed by economists which provide useful aids to thought about the difficult problems which arise in the field of energy economics.

Chapter 2 provides some statistical information about patterns of energy production and consumption and some discussion of the size of conventional energy reserves. In Chapter 3 the problem of resource depletion is considered both from a positive and a normative perspective. The choice of 'optimal' pricing policies for the energy industries is considered in Chapter 4. Chapter 5 confronts the difficult question of environmental externalities, examines the main sources of pollution, outlines the theoretical apparatus developed to cope with external costs and discusses the policy options open to governments and the implications for energy prices. Chapter 6 takes up the issue of tax policy towards the energy industries, concentrating particularly on the effects of various tax measures on depletion rates and the profitability of minerals extraction. The system of petroleum taxation in the U.K. sector of the North Sea is investigated in some detail and recent changes in the U.S. tax system are discussed with reference to the oil and natural gas shortage. Chapter 7 outlines some theoretical models of uncertainty and discusses what adjustments might be made in project appraisal to take uncertainty into account. We also comment on some of the ethical, political and philosophical problems which inevitably arise in this area. In Chapter 8 the technique of 'energy analysis' is described and criticised. The chapter as a whole differs from other chapters in that its concern with the economics of energy is indirect. However, a comparison of the two types of analysis is important since energy analysis has been advocated as a means of evaluating policy proposals. Finally, Chapter 9 provides a survey of U.K. energy policy since the early 1960s and of U.S. energy policy as presented in the *National Energy Plan* of 1977. In the last section of Chapter 9 we comment on the uses of economics in the formulation of a national energy policy.

# 2. STATISTICAL BACKGROUND

## 2.1 Introduction

The energy scene in the twentieth century both has been and will continue to be characterised by change. Thus, there have been fundamental changes in both the level and pattern of world energy production and consumption. New sources of energy have become commercially viable – for example, electricity from nuclear power stations – while other sources have increasingly encountered supply constraints, for example hydro-electricity in many developed countries. The consumption of both oil and natural gas has increased many-fold, and one of the principal concerns of energy policy in the late 1970s is the forecast of future shortages of oil.

The main purpose of this chapter is to provide a brief statistical sketch of some of the past and forecast changes in the energy scene which form a background to the discussion of energy policy. Apart from this introduction, the chapter is in five sections. The changing pattern of world energy consumption in this century is considered in Section 2.2. Some of the main trends in energy consumption in the United Kingdom and United States are considered in Section 2.3. Section 2.4 considers some of the changes and trends in energy production. Some data on conventional energy reserves are considered in Section 2.5, while Section 2.6 considers some of the non-conventional sources of energy.

## 2.2 The Changing Pattern of World Energy Consumption

The measurement of energy consumption poses a number of problems. A fundamental one is caused by the fact that different forms of primary energy are measured in different units.[1] Thus, coal is typically measured in tons while natural gas is measured in cubic metres. The adding-up problem for the different types of energy is usually solved by the use of a unit of account such as tons of coal equivalent, tons of oil equivalent, millions of barrels per day of oil equivalent, or therms.[2] The lack of consensus on the choice of a unit of account is compounded by a lack of agreement on how to convert from one unit of account to another. A major difficulty in the way of using standard units of account is that there are considerable variations in the calorific values of the various types of coal, crude oil and natural gas. In this chapter we have adopted a range of units of account. Some of the problems associated with the choice of *numéraire* are discussed is a different context in Chapter 8, Section 8.3.

Between 1925 and 1975 world energy consumption, measured in terms of oil equivalent, increased 5.8 times (see Table 2.1). During this period there was a marked change in the relative importance of the different fuels in the total energy mix. Although the consumption of coal doubled during this period it ceased to be 'the dominant source of energy, this role being taken by oil. Between 1950 and 1975 oil met 70 per cent of the additional requirements for energy. During this period there was also a major increase in the consumption of natural gas. As can be seen, by 1975 nuclear power was a relatively unimportant source of energy.

TABLE 2.1

*World Energy Consumption*

(mbdoe = million barrels a day oil equivalent)

|  | 1925 | | 1950 | | 1975 | |
|---|---|---|---|---|---|---|
|  | mbdoe | % of total | mbdoe | % of total | mbdoe | % of total |
| Hydro-electric | 0.5 | 2.4 | 2 | 5.4 | 7 | 5.8 |
| Nuclear | – | – | – | – | 2 | 1.6 |
| Natural gas | 0.5 | 2.4 | 3 | 8.1 | 21 | 17.4 |
| Oil | 3 | 14.3 | 11 | 29.7 | 56 | 46.3 |
| Solid fuel | 17 | 80.9 | 21 | 56.8 | 35 | 28.9 |
| Total | 21 | 100 | 37 | 100 | 121 | 100 |

SOURCE: 'Energy in Perspective', *Shell Briefing Service*, Dec 1977 (London: Shell International Petroleum Co., Shell Centre) chart 1.

While coal is a major source of energy most of it is consumed in the country where it is mined. In 1974 only 9 per cent of the world's production of coal moved in international trade.[3] This is in contrast to the situation for oil. Most of the major consuming countries have little oil of their own (the major exceptions being the United States and Russia) and have to obtain it from the major exporting countries. In consequence between 1950 and 1975 the growth in the international movement of oil was even greater than the increase in consumption. In the mid-1970s about 35 million barrels of crude oil were traded internationally every day.

Within this global picture there were significant country and regional differences in the consumption of energy. For example, over the period 1950 to 1975, coal consumption in the United States increased slightly, while in Western Europe it declined. During this period there was a substantial increase in the consumption of natural gas in both North America and Western Europe. In the mid-1970s in the Netherlands, gas from the large Groningen field supplied about 50 per cent of domestic energy consumption and was also exported to Belgium, France, Germany, Italy and Switzerland.[4] At that time the United Kingdom received 16 per cent of its total energy supply in the form of natural gas from North Sea fields.

Over the period 1950–75 Japan became a major consumer, and importer, of oil. In 1950 Japan was consuming 50 thousand barrels of oil each day and it ranked number 18 in the country rankings of total oil product demand. By 1975 this ranking had changed to number 2 (behind the United States) and Japan was consuming 4840 thousand barrels a day.[5]

Since 1950 oil has become an important source of chemical feedstocks. Approximately 8 per cent of total crude oil production in the non-Communist world is used for this purpose and 95 per cent of this areas's organic chemical production is derived from petroleum.[6]

## World Energy Supply and Demand to 2000

In recent years a number of studies have considered the future energy supply and demand situation for the non-

Communist world, with the particular objective of determining the date when world shortages of oil are likely.[7] These studies are unanimous in forecasting such shortages before 2000, and in two cases before 1985.[8]

Table 2.2 shows the forecast situation in 1985 for four of these studies. The principal reason for the lower energy consumption forecasts by the U.K. Department of Energy and the Workshop on Alternative Energy Strategies (W.A.E.S.) are their assumed lower economic growth rates compared with the O.E.C.D. Part of the reason for the C.I.A's higher demand forecast for OPEC oil relates to an assumed higher

TABLE 2.2

*Non-Communist World Energy Demand and Supply 1985 and Demand for OPEC Oil*

| | U.K. Dept of Energy (central case) | million barrels day oil equivalent | | |
| --- | --- | --- | --- | --- |
| | | O.E.C.D.[1] | C.I.A. | W.A.E.S. scenario D[3] |
| Assumed economic growth in O.E.C.D. area | 4.3% 1975−85 | 4.8% 1975−85 | 4.2%[2] 1975−85 | 3.1% 1977−85 |
| Energy consumption | 118.7 | 122.6 | | 109.5 |
| Energy supply (excluding OPEC oil) | 87.1 | 87.1 | | 75.9 |
| Net oil imports | 31.6 | 35.4 | 38.5−42.5 | 33.6 |
| Net Communist oil imports | − 0.8 | − 0.8 | 3.5− 4.3 | − |
| OPEC oil exports | 30.8 | 34.6 | 42−47 | 33.6 |
| OPEC oil consumption | 4.2 | 4.2 | 5−4 | 2.8 |
| Residual and increase in stocks | 0.8 | 0.5 | − | − |
| Demand for OPEC oil | 35.8 | 39.3 | 47−51 | 36.4 |

[1] O.E.C.D reference case.

[2] Excludes Australia and New Zealand.

[3] Low-growth growth-rate, oil at $11.5 a barrel and restrained national policies towards energy supply and conservation.

SOURCE: *Energy Policy: A Consultative Document*, Cmnd 7101 (London: H.M.S.O., 1977) table 3.2.

import demand from the less developed countries and from Communist countries.

To appreciate the significance of the 'demand for OPEC oil' forecasts they must be compared with forecasts of OPEC oil production capacity. In the mid-1970s this was about 39 million barrels of oil a day. Planned capacity for 1985 is probably in the range 45–47 million barrels a day.[9] Thus, in 1985 these forecasts suggest (apart from that of the C.I.A.) that there will be substantial spare capacity.

The forecasts contained in Table 2.2 assume unchanged energy policies by national governments. Various policies could be adopted to reduce the dependence on OPEC oil. The extent to which this dependence is reduced depends crucially on the policy of the United States, which is the world's largest energy consumer. The C.I.A. report forecast that U.S. oil imports could be as high as 16 million barrels a day by 1985. President Carter's energy proposals[10] were intended to limit these imports to 6 million barrels a day. The implementation of these proposals would have reduced the energy problems of other nations (see Chapter 9).

Looking further ahead to the year 2000 the W.A.E.S. report concluded that, even if energy prices rise by 50 per cent in real terms above their level in the mid-1970s, the demand for oil will exceed the supply. This will probably occur between 1985 and 1995. Because of long lead times alternative energy sources must be planned for now. The two most important alternative sources of energy in this century are likely to be coal and nuclear power. Coal reserves are relatively abundant (see Table 2.9 p. 21), but if extra supplies are to be required in the 1990s the required capacity must be installed in the 1980s (in the United Kingdom there is a time lag of about ten to twelve years between planning a new coal mine and its coming into operation – one reason for this relatively long time lag is the time taken up by public inquiries). Coal can be converted into synthetic natural gas (S.N.G.) and into liquid fuels. Although these conversions are, at the present time, uneconomic, by the 1990s a combination of higher oil prices and technical change could well make them commercially viable.

While electricity from nuclear power stations is capable of

making an important contribution to the world's energy supply by 2000, there is an important question of its public acceptability. This is especially important with regard to the construction of fast reactors.

## 2.3 Energy Consumption Trends in the United Kingdom and United States

The principal trends in energy consumption in the United Kingdom over the period 1950–75 are illustrated in Table 2.3. This period was characterised by a marked change in the energy mix, with a decline in coal consumption and an increase in the consumption of oil and natural gas. The substitution of these fuels for coal was the result of a com-bination of factors, such as a fall in their relative prices,[11] a change in consumer tastes in favour of the cleaner and more flexible fuels, and of various sectoral changes in the economy, such as the growth of road transport and the decline of the railways. During this period coal lost two of its traditional markets, gas-making and the railways. Natural gas from the North Sea has been substituted for manufactured town gas, and coal-fired steam locomotives have been replaced by diesel and electric locomotives. In 1950 there were approximately 20,000 coal-fired steam locomotives on the U.K. railways, and now there are none in regular service. The principal changes in the markets for coal in the United Kingdom are shown in Table 2.4.

Energy consumption in the United States is far higher than in any other country. In the mid-1970s it was consuming more than 30 per cent of the world's energy, although it had less than 6 per cent of the world's population. As can be seen from Figure 2.1 the United States uses more energy per dollar of G.N.P. than any other industrialised nation. It uses twice as much as West Germany, which has a similar standard of living. This high level of consumption has been stimulated by various governmental policies, such as the regulation of energy prices to consumers which kept their prices below the level of long-run marginal costs[12] and the payment of various tax benefits to producers.[13]

## TABLE 2.3

### U.K. Energy Consumption
(heat-supplied basis)

| | | billion therms | | | |
|---|---|---|---|---|---|
| | | 1950 | 1960 | 1970 | 1976 |
| *All classes of consumer* | | | | | |
| Solid fuel[1] | | 36.1 | 31.1 | 17.8 | 9.9 |
| Gas[2] | | 2.7 | 3.3 | 6.2 | 13.8 |
| Oil[3] | | 5.6 | 12.8 | 27.3 | 26.2 |
| Electricity | | 1.5 | 3.4 | 6.6 | 7.4 |
| | *Total* | 45.9 | 50.6 | 57.9 | 57.3 |
| *Energy consumption by source* | | | | | |
| *Industry* | | | | | |
| Solid fuel | | 15.7 | 13.9 | 8.9 | 5.3 |
| Gas[2] | | 0.9 | 1.5 | 1.9 | 6.1 |
| Oil[3] | | 1.7 | 4.5 | 11.4 | 8.6 |
| Electricity | | 0.7 | 1.5 | 2.5 | 2.8 |
| | *Total* | 18.9 | 21.4 | 24.7 | 22.8 |
| *Transport* | | | | | |
| Solid fuel | | 4.5 | 2.8 | 0.1 | – |
| Oil[3] | | 3.2 | 5.9 | 11.0 | 12.6 |
| Electricity | | 0.1 | 0.1 | 0.1 | 0.1 |
| | *Total* | 7.8 | 8.8 | 11.2 | 12.7 |
| *Domestic* | | | | | |
| Solid fuel | | 11.8 | 11.3 | 7.1 | 4.0 |
| Gas[4] | | 1.4 | 1.3 | 3.6 | 6.2 |
| Oil | | 0.2 | 0.7 | 1.3 | 1.4 |
| Electricity | | 0.5 | 1.2 | 2.6 | 2.9 |
| | *Total* | 13.9 | 14.5 | 14.6 | 14.5 |
| *Other* | | | | | |
| Solid fuel | | 4.1 | 3.1 | 1.7 | 0.7 |
| Gas[4] | | 0.4 | 0.5 | 0.7 | 1.5 |
| Oil | | 0.6 | 1.7 | 3.6 | 3.4 |
| Electricity | | 0.2 | 0.6 | 1.4 | 1.6 |
| | *Total* | 5.3 | 5.9 | 7.4 | 7.2 |

[1] Coal, coke, manufactured fuel.
[2] Town gas, natural gas and coke-oven gas used for non-energy purposes.
[3] Oil and creosote pitch mixtures.
[4] Town gas and manufactured gas.

SOURCES: *Digest of United Kingdom Energy Statistics* (London: H.M.S.O., various issues).

TABLE 2.4

*Coal Markets 1948–76*

|  | 1948 | | 1957 | 1970/1 | 1976/7 | |
|---|---|---|---|---|---|---|
|  | m. tons | % | m. tons | m. tons | m. tons | % |
| Power stations | 28.8 | 13.7 | 46.4 | 73.5 | 77.7 | 62.6 |
| Gas works | 24.6 | 11.7 | 26.4 | 3.5 | – | – |
| Coke ovens | 22.3 | 10.6 | 30.7 | 24.7 | 19.3 | 15.6 |
| Domestic | 36.4 | 17.3 | 35.6 | 18.4 | 10.4 | 8.4 |
| Industrial consumers | 38.0 | 18.1 | 37.5 | 18.5 | 9.1 | 7.3 |
| Collieries | 11.2 | 5.3 | 7.2 | 1.8 | – | – |
| Railways | 14.3 | 6.8 | 11.4 | 0.1 | – | – |
| Exports (plus bunkers) | 16.1 | 7.7 | 7.9 | 2.9 | 1.4 | 1.1 |
| Miscellaneous | 18.3 | 8.8 | 17.7 | 7.8 | 6.2 | 5.0 |
| *Total* | 210.0 | 100 | 220.8 | 151.2 | 124.1 | 100 |

SOURCE: *National Coal Board Annual Reports and Accounts* (London: National Coal Board, various reports).

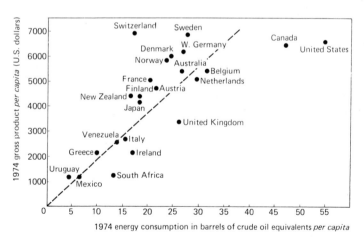

FIGURE 2.1    *Energy Consumption per Unit of G.N.P.*

SOURCE: *The National Energy Plan* (Washington: Executive Office of the President, The White House, 29 Apr 1977) p. 3.

TABLE 2.5

*Trends in Energy Consumption (United States)*

|  | 1920 | 1930 | 1940 | 1950 | 1960 | 1965 | 1970 |
|---|---|---|---|---|---|---|---|
| *Energy consumption* | | | | | | | |
| (trillion Btu) | 19,782 | 22,288 | 23,908 | 34,154 | 44,960 | 53,785 | 68,810 |
| (mbdoe) | 9.4 | 10.6 | 11.4 | 16.3 | 21.4 | 25.6 | 32.8 |
| *Energy consumption by source* (%) | | | | | | | |
| Coal | 78.4 | 61.2 | 52.4 | 37.8 | 23.2 | 23.0 | 20.0 |
| Natural gas | 4.2 | 8.8 | 11.4 | 18.0 | 28.2 | 29.9 | 32.8 |
| Oil | 13.5 | 26.5 | 32.4 | 39.5 | 44.6 | 43.2 | 43.0 |
| Hydro and Nuclear | 3.9 | 3.5 | 3.8 | 4.7 | 4.0 | 3.9 | 4.1 |

SOURCE: H. W. Richardson, *Economic Aspects of the Energy Crisis* (Farnborough, Hants: Lexington Books, Saxon House, 1975) table 1–1.

The main trends in energy consumption in the United States over the period 1920–70 are illustrated in Table 2.5. Over this period energy consumption increased by three and a half times, as compared with a more than fivefold increase in G.N.P. This increase was accompanied by significant changes in the energy mix. As in the United Kingdom, there has been a substantial decline in the relative importance of coal – from almost 80 per cent down to 20 per cent.[14] There has been a substitution of natural gas for coal in domestic and office heating, and by 1970, as in the United Kingdom, coal had ceased to be used for transportation with the phasing out of coal-fired steam locomotives. The decline in the relative share of coal has been offset by increases in the shares of oil and natural gas. These fuels have enjoyed a number of both price and non-price advantages over coal. They were generally cheaper, more flexible and associated with fewer environmental problems.

## 2.4  Energy Production

As would be expected the changes in energy consumption which were noted in Section 2.2 were accompanied by equally

TABLE 2.6

*Energy Production by Region 1950–75*

| Region | Oil (millions b/d) | Coal (millions metric tons) | Natural gas ($10^9$ cubic metres) | Hydro-electricity ($10^9$ kWh) | Nuclear electricity ($10^9$ kWh) |
|---|---|---|---|---|---|
| North America | | | | | |
| 1950 | 6.2 | 520 | 170 | 160 | – |
| 1975 | 12.5 | 570 | 620 | 520 | 180 |
| Caribbean and South America | | | | | |
| 1950 | 1.8 | 5 | 2 | 10 | – |
| 1975 | 3.6 | 10 | 25 | 100 | 2 |
| Western Europe | | | | | |
| 1950 | 0.1 | 490 | 1 | 110 | – |
| 1975 | 0.6 | 270 | 175 | 390 | 105 |
| Middle East | | | | | |
| 1950 | 1.8 | 5 | – | negl. | – |
| 1975 | 19.7 | 10 | 20 | 10 | – |
| Africa | | | | | |
| 1950 | negl. | 30 | – | 1 | – |
| 1975 | 5.1 | 70 | 5 | 35 | – |
| Far East and Australasia | | | | | |
| 1950 | 0.2 | 130 | negl. | 45 | – |
| 1975 | 2.2 | 200 | 25 | 150 | 20 |
| *Total* | | | | | |
| 1950 | 10.1 | 1180 | 173 | 326 | – |
| 1975 | 43.7 | 1130 | 870 | 1205 | 307 |

SOURCE: 'A Generation of Change: World Energy Patterns 1950–1975', *Shell Briefing Service,* Apr. 1976 (London: Shell International Petroleum Co., Shell Centre).

fundamental changes in the pattern of energy production. Table 2.6 shows that between 1950 and 1975 the pattern of energy production changed significantly between various regions of the non-communist world.

In Western Europe the period 1950–75 was characterised by the run-down of the coal industry and the substantial growth of the natural-gas industry. In Great Britain, for example, the coal industry in the post-war period has been

TABLE 2.7

*Trends in Output and Employment in the British Coal Industry*
*(National Coal Board mines)*

|  | 1947 | 1950 | 1960 | 1970 | 1976 |
|---|---|---|---|---|---|
| National Coal Board employment (000) | 710.5 | 687.9 | 587.5 | 285.1 | 241.6 |
| Number of collieries | 980 | 901 | 698 | 293 | 239 |
| Total output (m. tons) | 184.4 | 202.4 | 183.8 | 133.8 | 107.9 |

SOURCES: *Digest of United Kingdom Energy Statistics* (London: H.M.S.O., various issues).

characterised by a substantial decline in output, the number of collieries and employment, as is shown by the data in Table 2.7. Over the last three decades employment in the industry has declined by 66 per cent, 76 per cent of collieries have been closed and between 1950 and 1976 output fell by 53 per cent.

The growth of the natural gas industry and the substitution of natural gas for manufactured gas in the United Kingdom is shown by the data in Table 2.8.

TABLE 2.8

*Gas: Public Supply System in the United Kingdom*

|  | million therms | | | |
|---|---|---|---|---|
|  | 1966 | 1973 | 1975 | 1976 |
| Total gas available | 4,067 | 11,908 | 14,032 | 14,542 |
| North Sea gas | — | 10,630 | 13,367 | 14,030 |
| Total sales of gas | 3,685 | 10,729 | 13,112 | 13,997 |
| (a) Town gas | 3,685 | 2,323 | 713 | 212 |
| (b) Natural gas | — | 8,406 | 12,399 | 13,785 |

SOURCE: *United Kingdom Energy Statistics 1977* (London: Department of Energy, 1977).

In effect the period 1966–76 saw the phasing out of the manufactured gas industry and its replacement by a new natural gas industry.

## 2.5 Energy Reserves

The measurement of energy reserves for those energy resources which exist as stocks rather than as flows (e.g. solar energy) poses many difficult problems. However, an indication of these reserves is indispensable to the formulation of energy policy. Ideally the economist would like to have information on reserves presented in the form of a schedule indicating the amount suppliers would like to supply at different prices. Most available estimates of energy reserves, however, are not based upon specific assumptions about future prices and recovery technology. For fossil fuels, estimates of reserves are usually presented in two categories, proven and ultimately recoverable reserves. Proven reserves of oil, coal, etc., are usually defined as the amount of the resource which is recoverable from known reserves at today's prices and technology.

Ultimately, recoverable, or inferred, reserves, are an estimate of how much of the resource will eventually be produced. They allow for both new discoveries and changes in extraction technology. Before we consider some recent estimates of fuel reserves we will consider briefly some of the reasons why they have to be interpreted cautiously.

A fundamental difficulty for the measurement of energy reserves concerns the general conception of a fixed and known quantity of energy reserves.[15] There are a number of reasons for this. One is that the resources which are referred to as 'energy' are a function of the state of technology. For example, it is only with the development of commercial nuclear reactors in the 1950s that stocks of uranium have been classified as energy reserves. A second reason is that, for the calculation of energy reserves in addition to knowing what resources are in place, it is also necessary to know the proportion that is accessible; the recovery rate for these accessible reserves; and the extent to which it will be profitable to exploit the accessible unconventional sources of fossil fuels, such as tar sands.

There are many problems involved in determining the proportion of reserves which are accessible. Different methods of calculation may be used and both geological uncertainties and commercial considerations often lead to the making of conservative estimates.[16] Thus, it is not surprising that over time the reserve estimates for fossil fuels have typically been revised upwards. A good example of this is provided by estimates of world proved reserves of oil. Between 1947 and 1972 these increased nearly tenfold, from 9478 million metric tons to 91,376 million metric tons.[17] Part of this increase was due to new discoveries and extensions to known oilfields, but much of it was due to revisions of earlier estimates.

The recovery rate for accessible resources is a function of a number of factors, including the state of technology and geological conditions. The rate varies over time and between different deposits. In the United States, for example, the recovery rate for oil improved from 25 per cent in the 1940s to about 32 per cent by 1975.[18] At the present time, over the world as a whole, the recovery rate for oil averages 30 per cent; in the British sector of the North Sea recovery ratios of up to 40 per cent are being achieved.

There are various known sources of so-called unconventional sources of fossil fuels, such as the Orinoco tar sands in Venezuela, which, while accessible, have not been profitable to exploit at the ruling level of energy prices. However, as a result of the increase in oil prices since October 1973, the Shell Company announced in 1977 that it was going to invest between $3500 million and $4000 million in the extraction of oil from the Athabasca tar sands deposit in Alberta, Canada.

Some recent estimates of world proved and ultimately recoverable reserves of fossil fuels in 1975, along with recent estimates of uranium reserves, are presented in Table 2.9.[19]

These estimates show that oil is the fossil fuel nearest to exhaustion. At current rates of consumption proved reserves are sufficient for thirty years. By historic standards this figure is not especially low. In the post-Second World War period oil reserves calculated on a number of years' supply

TABLE 2.9

*Total World Fossil Fuel and Uranium Reserves and Consumption*
(mtoe = million tonnes oil equivalent)

| Fossil fuels | Proved reserves | Estimated ultimately recoverable reserves | 1975 consumption | Proved reserves ÷ 1975 consumption | Ultimate reserves ÷ 1975 consumption | Duration of ultimate reserves at 4% exponential growth rate |
|---|---|---|---|---|---|---|
| | thousand mtoe | thousand mtoe | thousand mtoe | years | years | |
| Oil | 80.4 | 233 | 2.7 | 30 | 90 | 37 |
| Gas | 56.5 | 171 | 1.1 | 50 | 155 | 51 |
| Coal | 329 | 645[1] | 1.9 | 175 | 340[1] | 71[1] |
| | | 3225[2] | | | 1700[2] | 110[2] |
| *Uranium*[3] | | | | | | |
| Up to 15 $/lb[4] | 19 | 37 | | | | |
| Up to 30 $/lb[4] | 32 | 59 | | | | |
| Up to 30 $/lb in fast reactors | 1590 | 2932 | | | | |

[1] and [2] 10 per cent and 50 per cent recovery rates respectively. There are no published estimates of recovery factors applicable to ultimate reserves of coal. There is a wide possibility of recovery rates, depending on economic and other factors. A range is therefore shown.
[3] Excluding Communist countries.
[4] Use in current designs of thermal reactors.

SOURCE: *Energy Policy – A Consultative Document*, Cmnd 7101 (London: H.M.S.O., 1978) table 3.3, p. 11.

basis have varied from a low of twenty-two years in 1947 to a peak of thirty-seven years in 1960. Major uncertainties relate to the level of undiscovered reserves and the rate at which they can be discovered and exploited, and there is certainly no consensus among oil experts on the size of world oil reserves.[20] As we have previously noted, using figures such as those given in Table 2.9, a number of recent reports[21] have concluded that the most likely future scenario for oil is that oil supplies will begin to level off in the late 1980s, reach a peak in the 1990s and decline thereafter.

It has been estimated that about 80 per cent of the proven oil reserves in non-Communist countries are in OPEC countries (61.6 thousand million tonnes out of a total of 76.0 thousand million tonnes) with 27 per cent in Saudi Arabia (20.8 thousand million tonnes).[22]

Compared with the Saudi Arabian reserves those of the United Kingdom are relatively small. The Department of Energy's best estimate is that total recoverable oil reserves in the United Kingdom will fall in the range 3–4.5 thousand million tonnes. These estimates allow for possible finds in areas which have not yet been licensed. With the given state of technology it is estimated that the upper limit for reserves in the present licensed areas is 3.2 thousand million tonnes (see Table 2.10). Measured in terms of the United Kingdom 1975 consumption of oil this latter reserve figure is equivalent to approximately 30 years' supply.

This figure is relatively high compared with the situation in the United States. Since about 1960 domestic oil consumption in the United States has outpaced domestic oil discoveries (excluding the discoveries in Alaska). In 1940 proved reserves in the United States were sufficient for fourteen years of consumption; by 1976 this figure had fallen to five years. Since the mid 1960s the United States has been dependent on oil imports.

## 2.6 Non-Conventional Energy Sources

The previous section was concerned with what are usually called 'conventional energy sources'. The various predictions

TABLE 2.10

*Estimated Oil Reserves In the United Kingdom Licensed Area*

| | million tonnes | | | |
|---|---|---|---|---|
| | Proven[1] | Probable[2] | Possible[3] | Possible total |
| Fields in production or under development | 1070 | 110 | 80 | 1260 |
| Other significant discoveries not yet fully appraised | 310 | 460 | 470 | 1240 |
| *Total present discoveries* | 1380 | 570 | 550 | 2500 |
| Expected discoveries — present licences (incl. 5th round) | – | 350 | 350 | 700 |
| Total present licensed areas (incl. 5th round) | 1380 | 920 | 900 | 3200 |

[1] Proven – those reserves which, on available evidence, are virtually certain to be technically and economically producible.

[2] Probable – those reserves which are estimated to have better than 50 per cent chance of being technically and economically producible.

[3] Possible – those reserves which, at present, are estimated to have a significant but less than 50 per cent chance of being technically and economically producible.

SOURCE: *Energy Policy – A Consultative Document*, Cmnd 7101 (London: H.M.S.O., 1978) p. 34.

which have been made about the increasing scarcity and high prices of these resources have in recent years increasingly focused attention on what are termed 'non-conventional energy sources'. There is no precise definition of these sources, but they are usually understood to include heavy oil and oil from oil shales and tar sands, geothermal energy, hydro-electricity, solar energy, wind and wave power. Many energy experts anticipate that these resources will make an increasingly important contribution to the world's energy requirements in the first half of the twenty-first century.

Because conventional oil has been relatively abundant and

inexpensive little attention has been paid to accessible deposits of oil sands, oil shale and heavy oil. Yet it is estimated that these deposits exceed the world's proven reserves of conventional oils. The development of these resources is associated with significant environmental problems (such as the disposal of the spent shale) and they have capital and operating costs which are higher than for conventional sources of oil. Estimates prepared by Shell International in 1977 suggest that the cost of oil from oil sands or shale in North America would probably lie in the range $15–25 (1976 U.S. dollars) per barrel of oil.[23] The W.A.E.S. report estimated that the maximum production from these sources in the year 2000 would be 3 million b/doe, mostly in North America.[24]

Geothermal energy is currently being exploited in a number of countries, including the United States, Italy and Japan. In 1974 there was 1400 MW(e)[25] of installed geothermal electricity generating capacity, representing an insignificant amount of energy on a global scale. This source of energy can be associated with significant adverse environmental effects (see Chapter 5, Section 5.2). In the United Kingdom geothermal energy is likely to be confined to a source of low-grade heat for heating purposes. The W.A.E.S. report did not expect this to be a significant source of energy by 2000.[26]

Most of the sites for hydro-electricity in the developed countries have already been exploited. In the developing countries, however, there is considerable potential. Although these countries have 44 per cent of the world's hydraulic potential, only 4 per cent has been exploited. The W.A.E.S. report concluded that hydro-electric supply in these countries could rise from 1 million b/doe in 1972 to 4.4 million b/doe in 2000.[27]

Solar sources of energy can be either direct in the form of radiant energy from the sun or indirect in the form of energy from the wind and waves. Solar energy systems involve high capital and low running costs, and thus their relative attractiveness is very sensitive to the size of the chosen discount rate. Solar energy does appear to have great potential. For example, it has been estimated that it could contribute 25 per cent of total U.S. primary energy by 2020.[28] The best-known application of solar energy is for water-heating.

Sufficient solar energy falls on the roof of the average house in North America, Europe and Japan to supply all of its energy requirements. There is, however, a storage problem since solar energy in many countries, such as the United Kingdom, is least plentiful in the winter when heating demands are greatest. At the present time, solar sources of energy are high-cost (partly because of the small production runs for the necessary equipment). Shell International have estimated that in 1976 solar hot water (on site at 35° latitude) cost over $40 (1976 U.S. dollars) per barrel of oil equivalent.[29] For the efficient use of solar energy, buildings need to be carefully sited, well insulated, and provided with adequate energy storage facilities. The W.A.E.S. report noted that since the turnover time of a nation's housing stock is approximately 100 years, progress with the use of solar heating would be slow if it was restricted to new buildings.[20]

There are various ways in which renewable energy can be harnessed from the seas. At the present time the only energy which is extracted from this source uses the tides. Tidal-power schemes are in operation at La Rance in France (240 MW(e)) and at Kislaya Guba in Soviet Russia. The Severn Estuary in the United Kingdom has been considered as the site for such a scheme.[31] In the United Kingdom an increasing amount of research is being undertaken into the harnessing of wave energy, since this is believed to be an important potential source of energy to the United Kingdom. In 1976 the U.K. Department of Energy selected four wave-energy projects for a £1 million research and development programme. In 1977 the budget was increased to £2.5 million for two to three years to allow large-scale work and field trials of two of the projects.[32] A major problem associated with such schemes is their potential effects on ocean shipping. It is unlikely that very much energy will be made available from this source by the end of this century.

Wind energy can be converted into either useful heat or electricity.[33] Because it is intermittent, effective storage is required. The electricity output from large-scale wind-driven generators is small. There is probably more potential for small-scale uses of this source of energy.

It has been estimated that it would be technically feasible to produce 30—40 million tons coal equivalent (mtce) from renewable sources of energy in the United Kingdom by 2000. However, in practice it is expected that the contribution will be a maximum of 10 mtce.[34]

# 3. THE DEPLETION OF ENERGY RESOURCES

## 3.1 Introduction

The onset of the so-called 'energy crisis' in the early 1970s stimulated great interest in the problem of resource depletion. Such interest is not a new phenomenon[1] although its mode of expression in computer model-building clearly is.[2] At its simplest the problem is obvious. Man lives on a planet with finite 'non-renewable' resources, resources which for several hundred years he has exploited at an increasing rate. With exponentially increasing demands on finite stocks a time will inevitably arise when they simply run out. Indeed with exponential growth that day can often appear uncomfortably close. Considerable debate has taken place on how large the world's reserves of resources are and, as we have seen in Chapter 2, estimates are inevitably tentative,[3] but in the context of exponential growth it would not appear to make much difference.

From the economist's point of view this type of debate always appears unsatisfactory. The idea that exponential trends will continue indefinitely without the price mechanism imposing some check seems counter-intuitive, while the exercise of measuring 'resources' in isolation from the state of technology and the costs of exploitation appears meaningless. This does not mean that the apocalypse *cannot* arrive but simply that at the very least powerful forces will assist to delay it. The two major factors preventing catastrophe are substitution possibilities and technical change.

(*a*) Substitution between resources while not solving the problem of a finite total stock nevertheless enables the economy to function without reliance on any *single* resource. In the field of energy the history of the last fifty years has been mainly concerned with the substitution of oil for coal.

Substitution in consumption, between less and more energy-intensive products (for example, public for private transport) would be a predictable outcome of rising energy prices, while substitution possibilities in production exist between non-renewable energy resources and other factors (the use of labour or capital inputs to conserve energy).

(b) The possibilities of technical change have been emphasised by many writers.[4] First, and most dramatically, technical developments can effectively create energy resources where none existed before. The wood shortage of Elizabethan England was finally resolved by the use of coal in the production of iron. Oil as a major energy source was unknown until the turn of the twentieth century. More recently uranium has become an important source of energy while in the future perhaps even deuterium (used in the process of nuclear fusion) will be added to the world's stock of depletable resources. Second, technical advances clearly play an important role in reducing the costs of extracting existing resources. Barnett and Morse,[5] for example, have traced the history of extraction costs and prices for major resources over nearly a century and found a decline in all but forestry. These cost reductions can occur in spite of recourse to resources of lower quality. Thus the story at least until the early 1960s seems to have been one of improved technology outweighing the tendency to diminishing returns. As for the future, the 'optimists' point to the possibilities of exploiting oil resources locked in shales and tar sands and to the development of processes such as coal liquefaction. Finally, technical change can enormously reduce resource input for a given output. Modern blast furnaces, for example, consume under one ton of coke for one ton of pig iron compared with over five tons of coke in the early nineteenth century.

In view of these possibilities of technical advance and factor substitution the concern with 'finite resources' and depletion rates might appear exaggerated. However, there are two major factors which help to explain it – uncertainty and pollution.

At any given time mankind lives with a given technology. The fact that, in the past, technical developments have enabled resource shortages to be overcome is in itself no guarantee that they will do so in the future. Thus to blithely assume that the benign influence of technical development will always come to the rescue can be seen as something of a gamble, and when 'doom' is involved, a gamble with high stakes.[6] Further, energy use is inextricably tied up with the question of pollution. Exhaust emissions from motor-cars, particulate emissions from coal-burning, radio-active waste disposal with nuclear power stations are all examples of serious pollution problems stemming from energy use (see Chapter 5). Such pollution is accompanied by health hazards, physical discomfort, and possibly even long-term climatic changes, although the latter appears to be very speculative.

Resource depletion as a policy issue is evidently very complex. Ideally optimal depletion rates would, given specified social objectives, be determined simultaneously with a policy towards capital accumulation, research and development, population size and pollution. In the following sections, however, we largely ignore the last three issues.

(*a*) Technical change is not only notoriously difficult to model but, as has been observed, resource-augmenting technical progress is sufficient to mitigate the problem of exhaustion. Thus the assumption of zero technical progress can be regarded as a useful device to focus attention on the most pessimistic case.

(*b*) The assumption of a constant population might be considered rather 'optimistic' in contrast to the 'pessimism' of (*a*). However with an exponentially growing population and zero technical progress 'doom' would indeed appear inevitable. If population is growing continuously at a constant rate, *some* resource augmenting technical progress is required to prevent the onset of declining consumption per head. Capital accumulation alone cannot do the trick if resources are 'essential' to production.[7]

(*c*) No attempt is made to specifically link the depletion rate problem with the treatment of pollution. The assumption is made that measures designed to control pollution exist. Although these may affect depletion rates (e.g. via the return on various types of investment) they are never overtly considered in the analysis.

The remainder of this chapter is in five sections. Section 3.2 outlines the traditional theory of inter-temporal choice and attempts to clarify some of the issues which arise concerning the discount rate. In Section 3.3 we outline the theory of exhaustible resources and discuss how the market would allocate a finite stock of resources over time. Section 3.4 introduces a more normative content into the analysis by outlining the implications of various social objectives. This section presents the opportunity of focusing attention on the important question of the substitutability between resources and reproducible capital. In Section 3.5 we specifically investigate sources of market failure in order to gauge whether, given the efficiency objective, resources are likely to be depleted too fast. Section 3.6 presents some brief conclusions.

## 3.2 Inter-temporal Choice

The problem of the individual consumer allocating his income over time can be illustrated in the standard two-period diagram, Figure 3.1. We assume that the individual concerned receives an income of $C_0^*$ in period zero and $C_1^*$ in period one. This configuration of receipts permits various opportunities for consumption within the area $0AB$. He may, for example, consume nothing in period zero and invest his income at the

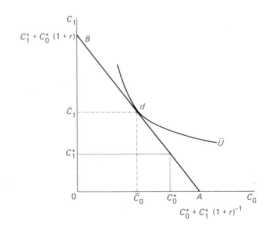

FIGURE 3.1

prevailing rate of interest $r$, thereby enabling the consumption of $C_1^* + C_0^* (1 + r)$ in period 1. Alternatively he might borrow as much as possible, again at the interest rate $r$, against his future income. The future income $C_1^*$ would therefore have to permit repayment of the principal $(\theta)$ and interest $(\theta r)$. Hence $\theta + \theta r = C_1^*$ or $\theta = C_1^* (1 + r)^{-1}$. In between these extremes are the possibilities along the line $AB$. This simply represents for each value of consumption in year zero, the maximum consumption in year one that the individual could attain *given* his income stream $C_0^*, C_1^*$. The present value of all the consumption possibilities along $AB$ is constant and equal to the present value of the consumer's income stream (i.e. his wealth). It is a simple matter to verify that the slope of $AB$ is given by $-(1 + r)$.

Which point on $AB$ the consumer chooses will depend on his preferences regarding present and future consumption. It is assumed that each individual has a utility function of the form

$$U = U(C_0, C_1),\tag{3.1}$$

which has the necessary properties to yield indifference curves such as $\bar{U}$ in Figure 3.1. In this case, therefore the highest utility attainable is at $d$, the individual consuming $\bar{C}_0$ in period zero and $\bar{C}_1$ in period one. This particular individual lends $\bar{C}_0^* - \bar{C}_0$, and receives in period one an extra $(\bar{C}_0^* - \bar{C}_0)$ $(1 + r)$ in addition to his income.

In equilibrium it is seen from Figure 3.1 that the slope of the individual's wealth constraint $-(1 + r)$ is equal to the slope of the indifference curve. The slope of the indifference curve at any point represents the marginal rate of substitution of $C_0$ for $C_1$ ($\mathrm{MRS}_{C_1, C_0}$), i.e. it represents the amount of consumption the individual would be prepared to sacrifice in period one for one extra unit of consumption in period zero. Although mathematically *negative* the $\mathrm{MRS}_{C_1, C_0}$ is usually taken as being *positive*. Hence it is found that in equilibrium

$$\mathrm{MRS}_{C_1, C_0} = (1 + r)$$

or $\qquad \mathrm{MRS}_{C_1, C_0} - 1 = r.\tag{3.2}$

The left-hand side of (3.2) is referred to as the individual's 'marginal rate of time preference' and in equilibrium this quantity equals the rate of interest.

The introduction of production possibilities into the analysis is shown in the 'Fisher Diagram' (Figure 3.2). The interpretation of $A^1B^1$ is exactly as in the simpler case, but now wealth is not merely the present value of $C_0^*$ and $C_1^*$. It also incorporates the production possibilities open to the individual as indicated by the curve $PP^1$. Our individual has the problem not simply of deciding upon combinations of consumption levels in the two periods *given* his wealth, but also of maximising his wealth *given* the production possibilities and other income. This problem is solved by the individual operating at point $Z$ on $PP^1$, investing $C_0^*C_0^p$ in period zero in order to achieve the increased income $C_1^*C_1^p$ in period one. The present value of the individual's wealth is maximised here as can be seen by the fact that the line $A^1B^1$ is as far as possible from the origin while still satisfying the production constraint $PP^1$.

The slope of $PP^1$ represents the marginal rate of transformation between present and future goods ($\text{MRT}_{C_1, C_0}$). At each point it shows the consumption that must be sacrificed in period one if consumption is to be one unit higher in period zero. Again it is seen that the wealth-maximising solution

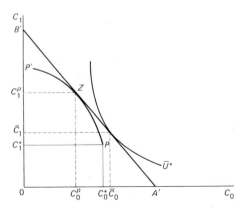

FIGURE 3.2

implies that

$$MRT_{C_1, C_0} = (1 + r)$$

or $\quad MRT_{C_1, C_0} - 1 = r.$ $\qquad\qquad$ (3.3)

The left-hand side of (3.3) is referred to as the 'rate of return in production' and in equilibrium it is equal to the rate of interest.

For the individual represented in Figure 3.2 the consumption combination $\bar{C}_0 \bar{C}_1$ now implies borrowing in the market. The individual borrows $\bar{C}_0 \bar{C}_0^p$ and repays $\bar{C}_1 \bar{C}_1^p$ in period one. $C_0^* C_0^p$ of these borrowings are used for investment purposes and $C_0^* \bar{C}_0$ to finance additional consumption in period zero.
As in the case of the usual static model of resource allocation the inter-temporal model (Figure 3.2) here will result in a Pareto-efficient allocation of resources providing that all the necessary conditions are fulfilled.

A Pareto-efficient configuration of the economy is defined as one which it is impossible to change without making someone worse off in their own estimation. It is based on the Paretian value judgements that (i) individuals are the best judges of their own welfare and thus if an individual prefers $A$ to be $B$ he is better off with $A$ than with $B$, and (ii) comparing two configurations of the economy, $C$ and $D$, $C$ is preferred to $D$ if at least one individual is better off in $C$ and no one is worse off. This value judgement avoids the need to make interpersonal comparisons. In addition to assuming given and constant tastes, population and technology, the following analysis assumes that all factors and products are perfectly divisible and enter into all relevant utility and production functions; the absence of external effects in production and consumption; certainty and perfect knowledge; and that all markets are perfectly competitive and in equilibrium.

Based on these assumptions necessary conditions for efficiency include the requirements that:

(1) For all individuals $MRS_{C_1, C_0}$ is the same. If this condition did not hold, borrowing and lending opportunities would exist which would permit a Pareto improvement in welfare. Suppose, for example, that for individual $A$

$MRS^A_{C_1, C_0}$ is 1.1, whereas for individual $B$ $MRS^B_{C_1, C_0}$ is 1.2. Clearly Mr $A$ could lend Mr $B$ a unit of consumption and be compensated as long as he received 1.1 units in the next period. Mr $B$, on the other hand, would be prepared to accept such an offer provided that he did not have to return more than 1.2 units in the next period. Any arrangement between these limits makes both individuals better off.

(2) For all individuals or firms $MRT_{C_1, C_0}$ is the same. The reasoning here is symmetrical with that above. If in one firm 0.9 unit of present consumption is sacrificed for an extra unit in the next period, whereas in another firm only 0.8 unit is sacrificed, there exists the possibility that, by transferring investment from the former to the latter, more could be consumed now with no consequent loss of consumption in the future.

(3) $MRS_{C_1, C_0} = MRT_{C_1, C_0}$. This is the 'higher-level' condition which is again best understood by imagining that it does not hold. If consumers are willing to sacrifice more present goods for an extra unit of future consumption than is required by production conditions then there clearly exists a possibility that a higher level of investment will enable everyone to become better off in their own estimation.

It can be seen from (2) and (3) that if all individuals face the same interest rate in the market then their own wealth and utility-maximising behaviour will lead to the attainment of these conditions. This result, however, depends crucially on the previously made assumptions, e.g. that all markets are perfectly competitive. It thus follows that Pareto efficiency in inter-temporal resource allocation will only result from market processes in a world of the imagination.[8] Some of the reasons for 'market failure' are considered in Section 3.5.

The rate of interest has so far been viewed as an exogenously determined factor to which individuals in the market adapt their behaviour. Like any other price, however, the rate of interest is itself determined by the forces of supply and demand. We have examined the factors which might underlie an individual's desire to lend or borrow – the fact that income may fluctuate over time or that productive investments may be undertaken. However, one important matter has still to be discussed – the issue of 'impatience'.

The degree of 'impatience' is usually referred to as the 'pure time-preference rate' or 'the rate of time-preference proper'. It refers to the rate at which individuals discount future *utility* not the rate at which in equilibrium they discount future consumption at the margin. Pure time preference is a characteristic of the individual's utility function and in terms of the usual indifference curve diagram it is measured by the marginal rate of time preference where consumption is equally divided between the two periods. Where there is no 'pure time preference' a consumption stream of (say) 10 units followed by 5 units should confer equal utility to the stream 5 units followed by 10 units. The first will not be preferred simply because the larger consumption basket occurs sooner. In these circumstances indifference curves will be symmetrical about the two axes and will have a slope of minus unity along a 45° line drawn from the origin, see Figure 3.3

Considerable discussion has occurred on the question of whether individuals 'ought' to discount future utility in this way. Ramsey and Pigou[9] both regarded such a procedure as irrational in the case of an individual and, if the utilities being discounted were those of individuals living in future periods, as morally unjustifiable. Our purpose in discussing this issue here is merely to highlight the possible differences in interpretation which can be given to a particular discount rate, differences which in the context of energy policy become rather significant. In particular it is often asserted that high

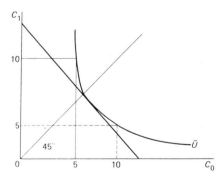

FIGURE 3.3

interest rates 'favour' the present over the future and vice versa.

It is as well at this point to distinguish between the discount rate as a policy instrument and the discount rate as a market-determined price. The use of a higher rate of discount to appraise a particular project will, all other things remaining constant, certainly tend to favour those whose benefits occur at an earlier date. On the other hand a rise in the market-determined rate of discount is much more difficult to interpret. In the perfectly competitive traditional textbook model the interest rate is merely an equilibrium price which emerges from the transactions of many individuals attempting to solve the inter-temporal maximisation problem sketched above. Just as price ratios enable rates of substitution and transformation to come into line in the static model, so in the inter-temporal case the interest rate performs the same function.

A rise in the market rate of interest will therefore *reflect* changes which have occurred either in production possibilities or in preferences. Whether these changes militate against future consumption cannot be inferred simply by looking at the change in the interest rate itself. As an example consider three possible interpretations of an observed increase in the interest rate in a hitherto perfect market.

(*a*) The rise in the interest rate is the result of the introduction of a proportional tax on the returns to investment. Individuals are less willing to lend for lower after-tax returns and an 'inefficient' gap now yawns between the marginal rate of transformation (the before-tax return) and the marginal rate of time preference (the after-tax return). The increased market rate of interest faced by the producer causes him to cancel investment projects while the lower returns to the lender induce him to substitute present for future consumption. We conclude that the new situation is:

(i)  Pareto inefficient,

and (ii) encourages present consumption at the expense of the future.

(*b*) The rise in the rate of interest stems from the success of a social movement which preaches that life becomes progressively more miserable with advancing age. Goods are less enjoyable at retirement than they are in youth. A section of the population begins to exhibit 'pure time preference' and decide to enjoy life while it is still possible. The higher interest rate is consequently:

(i) Pareto efficient,

and (ii) *by definition* encourages present consumption at the expense of the future.

(*c*) The rate of interest has risen because a new discovery has raised the productivity of capital and caused businessmen to compete for scarce funds to invest. The marginal rate of transformation between present and future goods has increased, and an increased rate of interest is necessary to achieve efficient intertemporal allocation in the new situation. A higher interest rate is therefore

(i) Pareto efficient,

and (ii) evidence that the future will be richer than the present.

It has seemed worth while to spell out the above possibilities since this issue is of considerable importance in the field of energy economics. As will be seen, the rate of discount enters into the depletion-rate decision in a particularly significant way and the question of whether prevailing rates are 'correct' is of central concern. Those theorists and others who are pessimistic and doubt the capacity of markets to provide for the future emphasise the possibilities of market failure and pure time preference (points (*a*) and (*b*) above). Those who, on the contrary, take a more optimistic view emphasise that the future will be better provided than the present with capital and that this future wealth is reflected in a positive discount rate.

## 3.3 Resource Depletion: Some Positive Theory

A number of elementary theorems of the 'pure theory of exhaustion' are outlined in this section. This theory attempts to deduce the behaviour of resource owners under conditions of certainty with respect to how fast they will deplete their stock.[10] We start with the simplest case in which a firm has the rights to a fixed stock of a homogeneous resource ($q$) which it can extract at a constant cost per unit ($C$). Consideration of this case leads to the 'fundamental principle of the economics of exhaustible resources'[11] – the market price of the resource net of extraction costs must rise at a rate equal to the rate of interest.

That the 'fundamental principle' must hold is perhaps most easily seen by imagining a case in which market price minus extraction cost rose at a rate lower than the interest rate. In such a case, a profit-maximising firm would extract and sell its whole stock as soon as possible, investing the proceeds in alternative areas which yield the market rate of interest. Any delay could only reduce the present value of the firm's profits. On the other hand a firm facing a 'net price' which is rising faster than the rate of interest would have every incentive to leave the resource in the ground since it would in effect represent a superior investment to any alternative. A net price rising at the rate of interest is therefore an equilibrium condition in the asset market as well as the market for output. Only such a trend in net price is compatible with a positive output in every period (firms will be indifferent as to the time at which they sell) while resource owners will, at each period, be just content to hold the stock available.

The model can be set up more formally as follows. Let profit in each period be represented by $\pi_t$. Then, in a competitive market,

$$\pi_t = p_t q_t - C q_t, \tag{3.1}$$

where $p_t$ = market price of resource, $q_t$ = output in period $t$,
$C$ = marginal cost of resource (assumed constant).

The firm will then attempt to maximise the discounted value of its profit subject to the constraint that

$$\sum_{t=0}^{T} q_t = \bar{q}, \tag{3.2}$$

where $\bar{q}$ is the given stock of resource, and $T$ is the time horizon which we here assume to be exogenously determined.

The final problem may therefore be written,

$$\text{Max} \sum_{t=0}^{T} \pi_t(1+r)^{-t} = \sum_{t=0}^{T} (p_t q_t - C q_t)(1+r)^{-t},$$

subject to (3.2) above. We now form the Lagrangean expression

$$L = \sum_{t=0}^{T} (p_t q_t - C q_t)(1+r)^{-t} - \lambda \left[ \sum_{t=0}^{T} q_t - \bar{q} \right]$$

Differentiating with respect to each $q_t$ we then obtain the $T$ first-order conditions

$$\frac{\partial L}{\partial q_t} = (p_t - C)(1+r)^{-t} = \lambda, \quad (t = 0, 1, \ldots, T).$$

or

$$p_t = C + \lambda(1+r)^t. \tag{3.3}$$

Price in period $t$ must equal marginal extraction costs *plus* an expression $\lambda(1+r)^t$. Scott[12] has suggested that this be termed a 'user cost' since it arises from the fact that using the resource in the present eliminates the possibility of its use in the future. Equation (3.3) then indicates that price in each period will equal marginal extraction cost plus user cost.

Writing $R_t = P_t - C$ we can now easily obtain

$$\frac{R_{t+1} - R_t}{R_t} = \frac{\Delta R_t}{R_t} = r, \tag{3.4}$$

which is, of course, the 'fundamental principle'. The term $R_t$ has been referred to above as 'net price', often it is called 'royalty', 'profit', or 'rent'. Probably 'royalty' or 'rent' come closest to accurately reflecting its nature. In a competitive industry with free entry it would be surprising to discover a condition which implied rising profits accruing to firms over time. In fact, of course, the profits represent returns to the owner of the resource deposits who can charge the competitive firms for the privilege of working them.

## The Depletion Time

In the preceding analysis the time of depletion $T$ was imposed by assumption on each firm. This is clearly unsatisfactory since presumably resource owners will have to determine the rate of output and the depletion time $T$ simultaneously. Of course, if it can be assumed that price may rise without limit, the date at which all stocks of the resource are exhausted may never be reached. Analysis usually proceeds, however, on the assumption that at some limiting market price the quantity of the resource demanded declines to zero, and it is this price together with the rest of the resource demand curve, total resource availability and the 'fundamental principle' that between them determine the path of industry output and market price over time.

The argument is best illustrated using figure 3.4.[13] The curve on the right-hand side represents the demand for the resource at various prices per unit. It is assumed that the upper limit on price is given by point $p^*$. A common justification for this is the idea that some alternative source of supply will become available when the price reaches a sufficiently high level. On the left-hand side the curve with the arrows ($ab$) represents the price at each point in time, a path which, given the demand curve, must imply a given path of output. The determination of curve $ab$ can be envisaged as a process of trial and error. Given a certain value of $\lambda$ a path similar to $ab$ will be determined. If before it reaches price $p^*$

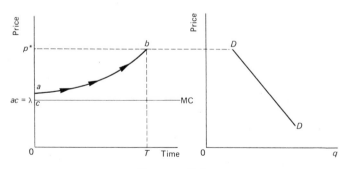

FIGURE 3.4

the stock of resources is exhausted there is clearly scope for a higher price at each point in time and resource owners will see the possibility of receiving higher royalties. If, on the other hand, resources remain unexploited at price $p^*$ the owners of the deposits will, in a competitive market, be under pressure to reduce their royalties. Eventually, some path *ab* will be discovered which obeys the 'fundamental principle' and tracks to the maximum price $p^*$ just when exhaustion occurs.[14]

It is worth noting that a path such as *ab* in Figure 3.4 is attained only because the future is assumed to be *certain*. The size of the resource stock is known, its content is homogeneous, the shape of the demand curve is known and, most important, there are futures markets which operate perfectly. Contracts are made *now* to deliver or purchase the resource at specified prices and dates. Thus although the figure shows a path of prices altering through time, they are all determined at time zero. The absence of such conditions in reality is clearly a major obstacle to such a smoothly operating market.

## Comparative Statics

It is now necessary to review briefly the effects upon the model of changing conditions. The most important of these include changes in the rate of interest, resource availability, costs of extraction and demand for resources.

### (a) Interest rates

The rate of interest is clearly of central importance in determining the path of prices over time. Suppose to begin with that the rate of interest falls. From Figure 3.5 it is seen that if the price of the resource at the outset remains unchanged at *a* the curve showing the time path of price in the new conditions would lie below the curve *ab*. However, if this time path actually materialised, stocks of resources would clearly be exhausted before price $p^*$ is reached. Just as resource prices cannot lie everywhere below *ab*, they equally cannot be everywhere above since in this case unexploited resources would remain. The conclusion therefore must be that a fall in the rate of interest raises the initial royalty and market price,

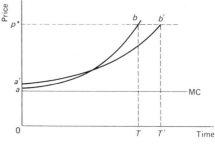

FIGURE 3.5

reduces the rate of increase in the royalty, but extends the date of depletion beyond $T$. In Figure 3.5 the time path might be represented by a curve such as $a^1 b^1$.

It should be recognised that while the above argument provides reasons to expect lower interest rates to slow down depletion rates, thus acting in a 'conserving' manner, the analysis fails to take into account changes in costs which might result from alterations in the rate of interest. Gordon[15] has examined this possibility and shown that where lower rates of interest result in lower capital costs there is no longer an assurance that depletion will be delayed.

### (b) Costs

As may be inferred from the preceding paragraph, a fall in marginal extraction costs will, *ceteris paribus*, lead to precisely the opposite results of a fall in the interest rate.[16] A decline in costs with the initial price unchanged will clearly lead to a rise in the initial royalty. This higher royalty rising at a rate equal to the interest rate will result in a time path of prices lying everywhere above the preceding path and hence in unexploited stocks of resources when price $p^*$ is reached. By a precisely parallel process of reasoning to case $a$ it must be concluded that market price will start out lower but finish higher after costs have fallen. The depletion time will therefore be brought forward. A converse argument establishes that a rise in extraction costs will lower the royalty, increase initial prices and lower later prices thereby reducing the quantity demanded in the earlier periods and increasing it later. This has an important bearing on tax policy. Since a tax

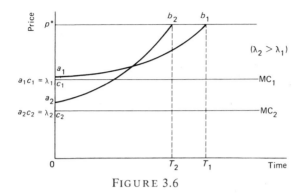

FIGURE 3.6

on each unit of the resource is equivalent to a rise in marginal costs, the above reasoning suggests that such a policy instrument will slow down the rate of depletion. These conclusions are summarised in Figure 3.6. Curve $a_1 b_1$ corresponds to cost conditions $MC_1$, while curve $a_2 b_2$ corresponds to cost conditions $MC_2$.

### (c) Additions to resource stocks
It is unnecessary to trace the argument in detail here, since the method is similar to that used in cases (a) and (b). If resource stocks rise the original time path will leave them unexploited. Royalties must therefore fall and the final result will be a path of prices everywhere below the original. More resources will be used at each point in time but resource exhaustion will nevertheless be delayed.

### (d) Changes in demand
Assuming that $p^*$, the price at which demand falls to zero, is unchanged a rise in demand will increase royalties and reduce the length of time to exhaustion.[17]

### Variable 'Quality'

Until this point the analysis has been greatly facilitated by the assumption that marginal costs of extraction are constant. In reality, of course, mineral deposits vary greatly in terms of mineral content and accessibility. The production of a barrel of oil may vary greatly in extraction cost not only between fields but between points in time in the same field.[18] The 'pure theory of exhaustion' as outlined above is therefore limited in its applicability, and although simple alternative rules to the 'fundamental principle' cannot be derived it is evident that incorporating differing cost conditions into the analysis can lead to conclusions quite distinct from those derived earlier. It turns out that the 'fundamental principle' does not go into abeyance in the more complex setting however, but operates in a more subtle way.

Suppose that there are two sources from which it is possible to obtain oil. This oil, it will be assumed, is of identical type and quality when produced but cost conditions vary between locations because of communications problems, geological differences, climatic conditions and so forth. The first thing to note is that, in the presence of differing costs, theory would predict that only one source would be exploited at any one time and that the lower-cost source would be exploited first. To simplify the problem as much as possible assume that marginal costs of production are constant at each location.[19] Marginal costs at the more accessible deposit are designated $C_1$ and costs at the less accessible deposit $C_2$ where $C_2 > C_1$. Suppose further that, as before, the extraction industry is competitive in both locations. It is very simple to show that, in these conditions, no single rate of price increase can simultaneously satisfy both sets of producers and resource owners.

For the owners of the low-cost deposits the net price of output must rise, as we have seen, at a rate equal to the interest rate if positive production is to occur in each period from this source and equilibrium is to prevail in asset and product markets. Assuming that this is the case, what would be the impact on resource owners of the high-cost deposit? For the low-cost source we know that:

$$(p_t - C_1) e^{-rt} = \lambda$$

or          $p_t - C_1 \qquad = \lambda e^{rt}.$

Hence     $\dfrac{d}{dt} (p_t - C_1) \quad = r\lambda e^{rt}$

and $\dfrac{\dfrac{d}{dt}(p_t - C_1)}{p_t - C_1} = r.$

This result is, of course, the equivalent of equation (3.4) only here expressed in terms of continuous time rather than discrete time intervals. The net price on the best deposits is increasing at the interest rate. In such a case, however, the net price on the poorer deposits would be given by:

$$p_t - C_2 = \lambda e^{rt} + C_1 - C_2.$$

Thus $\dfrac{d}{dt}(p_t - C_2) = r\lambda e^{rt}$

and $\dfrac{\dfrac{d}{dt}(p_t - C_2)}{p_t - C_2} = \dfrac{r\lambda e^{rt}}{\lambda e^{rt} - (C_2 - C_1)} > r.$ (3.5)

Since $(C_2 - C_1) > 0$ the above expression must be greater than $r$ because the denominator is less than $\lambda e^{rt}$. A net price rising at the rate of interest for the output from the low-cost deposit implies therefore a net price rising at a faster rate in the case of the high-cost deposit. In these circumstances, as we have seen, the owners of high-cost deposits would prefer to leave them unexploited.

The question then arises, what would happen if the price of oil rose in such a way that the royalty on high-cost deposits increased at the rate of interest? In such a case it is easily verified that the net price of low-cost deposits would rise according to equation (3.6).

$$\dfrac{\dfrac{d}{dt}(p_t - C_1)}{p_t - C_1} = \dfrac{r\lambda e^{rt}}{\lambda e^{rt} + (C_2 - C_1)} < r.$$ (3.6)

Thus the net price on low-cost deposits would be rising at a rate below the rate of interest. This situation is untenable, however, because owners of the low-cost resource would deplete as soon as possible rather than earn less than the rate of interest on their assets.

We conclude therefore that a perfectly functioning market must result in the lower-cost deposits being exploited first with the net price rising at the interest rate $r$ and owners of higher-cost resources being content to leave them untouched. At the point of exhaustion of the first deposits the poorer ones will then come into production again with the net price increasing at the rate of interest. This time path of prices is illustrated in Figure 3.7 once more on the assumption that $p^*$ represents a limit on the price of the resource. Low-cost deposits are exhausted at $T_1$ and the rest at $T_2$. Herfindahl,[20] who first introduced this analysis of the two-grade case, investigates the effect of varying the proportions of high-cost to low-cost sources on the time to depletion and the path of prices. Here we simply note two basic points:

(*a*) It is no longer true that the observed net price will, between any two dates, have grown at the rate of interest. The 'net price' at time $T_1^*$, for example, is $ab$, whereas at time $T_2^*$ it is $cd$. Measured between these two points in time therefore it might appear that the net price is falling.

(*b*) On the other hand it remains true that the net price on *the grade of resource currently being exploited* must be rising at the rate of interest.

(*c*) It is important to distinguish between the 'net price' of

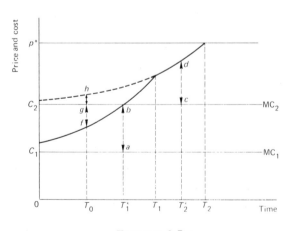

FIGURE 3.7

a resource in the *product market*, i.e. the difference between market price and extraction cost, at a given point in time, and the 'royalty' which may now be conveniently defined as the price of the resource deposit *in the asset market*. In the case of a homogeneous resource they tend to amount to the same thing. When we allow for differing grades of resource there are some subtle differences. The net price and the royalty as defined above will be the same only when the deposit is actually being worked. In the case of the high-cost grade in the example above, it has already been noted that 'net price' is rising at a faster rate than the rate of interest before time $T_1$. Indeed at time $T_0$ the 'net price' of this grade would be negative since the market price falls short of marginal extraction costs. It obviously does not follow, however, that the price of this grade *as an asset* to hold at time $T_0$ is negative. Clearly operators in the asset market will realise that after time $T_1$ the deposit may be worked at a profit and will be willing to pay a price or 'royalty' equal to the present value of those future profits. The asset value of the low-cost grade per unit at time $T_0$ is given by the distance *gh* while 'net price' is the negative distance *gf*. This distinction enables us to clarify the operation of the 'fundamental principle' in the case of differing extraction costs. Although the 'net price' of both resources cannot increase at the rate of interest it remains true that the 'royalty' or the asset value of both grades increases at the rate of interest.

## Rising Costs with Cumulative Extraction

The last observation carries over into the more general analysis of cost conditions. In principle we might imagine a case in which the number of deposits rises and all face different cost conditions. Indeed we might then go on to allow for varying grades and types of resource within a single deposit, arguing that it is possible to think of these grades conceptually as different deposits. And finally we might allow costs to rise continuously with cumulative output — perhaps as recourse is had to lower grades or less accessible deposits, or as pressure declines in an oil reservoir.

Although the analysis of such a model is more complex

than the simple two-grade example above the basic conclusions are unaltered. At each point the resource owner asks the question: given the future trend of prices and given the cost conditions will the net price of this marginal unit of resource increase at a faster rate than the rate of interest? If the answer is 'yes' he will leave it where it is. At each point of time, however, the rate of increase in the net price earned on the unit of resource currently being produced will be precisely equal to the rate of interest. Further the royalty or asset value of each unit of unworked resource will be continuously increasing at the rate of interest. The absolute size of the current observed net price no longer necessarily rises over time as we noted in the simple two-grade example. Indeed if costs of extraction can rise up to and above $p^*$ it is evident that the final units of production might sell at prices *equal* to extraction costs, thus reducing the net price to zero.[21]

In the case of a homogeneous resource produced under conditions of constant costs the importance of adequate futures markets to the attainment of the time path of prices and outputs was emphasised. Clearly the burden placed on this assumption is even greater in the context of varying extraction costs. Futures markets will have to operate in each separate grade of resource, and where costs rise with cumulative extraction each unit of output must conceptually be regarded as a separate grade of resource.[22]

## 3.4    Resource Depletion: Some Formal Normative Approaches

Thus far the analysis has concentrated on the question of how fast a perfectly competitive extractive industry will deplete an exhaustible resource. In this section we turn to the controversial question of how fast a given stock of exhaustible resources 'ought' to be depleted.

### (A)  *Pareto Efficiency*

In static analysis the notion of 'efficiency' gained in influence, partly at least because it appeared more 'scientific' than its predecessors. It specifically eschewed inter-personal comparison of utility and relied on what were thought to be rather weak value judgements – that individuals knew their own interests best, that individual valuations were those to appear in the 'social welfare function' and that the latter was specified in such a way that social welfare would be improved if at least one person could be made better-off in his own estimation without making another worse-off. Extending the criterion to the inter-temporal case poses substantial problems, practical and philosophical if not formal.

Adopting Pareto efficiency as a social objective in no way avoids the problem of deciding whose utilities are to count in determining an efficient state. In the static allocation case the problem, although still present, is more tractable. Usually it is asserted that everyone's utility is to count. In the inter-temporal case more difficulties arise. If literally *everyone*'s utility is to count, no doubt the model might be extended to incorporate this value judgement in a purely formal sense and sets of efficiency conditions could be derived which would characterise an efficient inter-temporal allocation.[23] But this would still leave to the imagination the task of envisaging institutions – markets or otherwise – which might have some chance of attaining these conditions.

The importance of perfectly functioning futures markets in the context of resource depletion has already been emphasised. The model of Section 3.2 with its assumption of perfect certainty requires that a society of individuals meets together at time zero with given factor endowments, well-established property rights and known production possibilities, to form contracts with one another to supply goods or factors at differing times in the future at market-clearing prices. All individuals must be assumed to know with certainty the state of technology now and in the future, as well as the size of the stock of any exhaustible resources. Where the society of individuals is to include generations yet unborn the difficulty of achieving efficiency, even where futures markets exist, is obvious. Future generations can bring no endowment of their own to operate in existing futures markets. Such markets may allocate a depletable resource efficiently among the members of a given generation, but there might seem little hope that the members of future generations will be considered.

This conclusion is, however, open to criticism.[24] Markets, it can be argued, may not behave quite as inefficiently from the point of view of future generations. Clearly owners of an exhaustible resource will not be unaware of a potential demand for its services in the distant future even if no contracts for future delivery have been obtained, and may calculate that higher prices in the future will make conservation worth while. Thus the future does influence present markets, not by bidding in them and confirming with certainty that the world will continue, but by the expectations of the future which are held by the present generation. Markets will exist in the future, and future generations will bring endowments to bid in them. Whether these endowments will be 'fair' relying as they do on the behaviour of the present is not a matter on which the Pareto criterion can enlighten us any more than it can about the distribution of income among individuals existing at a given time. The issue, as far as the efficiency criterion alone is concerned, then becomes one of deciding how close to the 'efficient' path an economy will track relying on accurate expectations rather than perfect futures markets.

An alternative approach to efficiency is to define it in terms of those individuals now existing and to ignore the utility of future generations, except in so far as knowledge of the latter has an effect on the wellbeing of the former. Defined in this way it can be shown[25] that a perfectly competitive industry in the presence of perfect futures markets will result in efficient inter-temporal allocation of an exhaustible resource providing that all other sources of market failure are absent. Ignoring the wellbeing of the future is a standpoint which will be morally unacceptable to many. Even those who argue that, if historical evidence is anything to go by, the future will be able to take care of itself, are not necessarily suggesting that future generations ought not to be considered, but rather are making optimistic assertions about their likely condition. In the following section therefore we briefly discuss the implications of some 'stronger' welfare criteria.

## (B)  *Utilitarian Objectives*

In Section (A) it was observed that by adopting efficiency as
an objective or criterion for evaluating inter-temporal allocation the analyst is effectively remaining 'neutral' about the
inter-temporal distribution of income. The future may be
rich relative to the present or it may be poor, but for 'efficiency' the only question is whether a different allocation of
inputs and outputs over time could make any person better-
off without making another person (living perhaps at another
time) worse-off. An alternative approach is to make quite
explicit value judgements about distribution.

Suppose that each person living at a given time receives
utility from consumption according to the function

$$U_i = U_i(C_i).$$

It is assumed that utility here is cardinally measurable and is
not simply a means of ranking alternatives applicable to a
particular individual as in usual consumer theory. Thus if
individual $A$ obtains three utils of satisfaction from a given
level of consumption and $B$ obtains four utils we are permitted
to deduce that individual $B$ is more satisfied than individual
$A$ and that the sum of satisfactions is $4 + 3 = 7$ utils.

Plausible properties are now asserted to characterise the
function $U_i(C_i)$. In particular $U_i'(C_i) > 0$ and $U_i''(C_i) < 0$. That
is, an addition to an individual's consumption always increases
his utility but at a diminishing rate. A sequence of important
assumptions then establishes the form of an inter-temporal
social welfare function (SWF).

(i)  All individuals are identical.

(ii)  Total social welfare at any one time is defined to be
the sum total of individual utilities.

(iii)  Total social welfare over all time is the discounted
sum of social welfare at each point in time.

For the purposes of analysing resource depletion some further
assumptions are usually made:

(iv)  Total population is the same in each period.

(v)  Governments are assumed to undertake any necessary

intra-temporal redistribution measures so that aggregate consumption at a given time is related to total welfare by $W = U(C_t)$. Since all individuals are identical we are left with an inter-temporal SWF of

$$W^* = \sum_{t=0}^{\infty} \delta^t U(C_t), \quad 0 \leqslant \delta = \frac{1}{1+r} \leqslant 1, \tag{3.7}$$

where $r$ is the rate of discount. We are now in the position to investigate an optimal consumption path for a non-renewable resource.

## The 'Cake-eating' Model

The simplest possible case is to imagine a situation in which a given stock of a consumption good $R$ exists, there are no production possibilities and the problem is to use the stock in such a way as to maximise (3.7). Thus

$$\text{Max} \quad W^* = \sum_{t=0}^{T} \delta^t U(R_t) \quad \text{subject to} \quad \sum_{t=0}^{T} R_t = \bar{R},$$

where $\bar{R}$ is the given stock of the resource.

We now form the Lagrangean

$$L = \sum_{t=0}^{T} \delta^t U(R_t) - \lambda^* \left[ \sum_{t=0}^{T} R_t - \bar{R} \right]$$

Hence

$$\frac{\partial L}{\partial R_t} = \delta^t U'(R_t) - \lambda^* = 0$$

or $\qquad U'(R_t) = \lambda^* \delta^{-t} = \lambda^* (1+r)^t. \tag{3.8a}$

Equation (3.8a) indicates that the marginal utility of consumption of $R$ should optimally increase at a rate equal to the discount rate.

i.e. $\qquad \dfrac{U'(R_{t+1}) - U'(R_t)}{U'(R_t)} = \dfrac{\lambda^* [(1+r)^{t+1} - (1+r)^t]}{\lambda^* (1+r)^t} = r.$

$$\tag{3.8b}$$

The implication of a rising marginal utility must be a declining consumption of $R$ over time. In the case where the moral strictures of Ramsey and Pigou are taken to heart and future utilities are not discounted ($\delta = 1$ or $r = 0$) equation (3.8a) tells us that the marginal utility of consumption must be constant over time, implying a constant level of consumption. The two time paths of consumption of $R$ are shown in Figure 3.8.

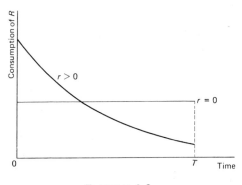

FIGURE 3.8

In the above example the period of time $T$ over which the allocation problem stretches is fixed by assumption. A more general case would be one in which the time horizon stretches to infinity. In fact it can be shown that the qualitative characteristics of the time path will be unaltered. Obvious problems arise, however, if positive quantities of good $R$ require to be consumed per unit of time in order to achieve subsistence. Clearly in these circumstances the stock cannot be made to last indefinitely, and the inter-temporal allocation problem involves fixing a date for the end of the world. As one would expect, the existence of pure time-preference advances this fateful day while a zero pure time-preference rate recommends a policy of equal consumption per period. Whether this equal consumption level implies continuous *subsistence* consumption (the so-called Inter-temporal Calcutta solution) depends upon the properties of the assumed

utility function $U(R_t)$. That it might *not* do so can be illustrated by considering the decision to move away from a pure
subsistence regime. By surviving for one period shorter, that
periods subsistence stock of $R$ can be redistributed equally
across the remaining periods. Suppose that there are $n$ remaining periods and suppose that the subsistence of stock of $R$ say
($S^*$) is divided equally among them. Clearly the utility gain
will be

$$nU'(R)\frac{S^*}{n} = U'(R)S^*,$$

while the utility loss will be $U(S^*)$.

The duration of survival should evidently be shortened as
long as

$$S^*U'(R) > U(S^*)$$

or      $$U'(R) > \frac{U(S^*)}{S^*},$$

i.e. as long as the marginal utility of consumption of $R$
exceeds the average utility gained from present consumption
levels.[26]

### Introducing Production

The 'cake-eating' model is of only limited interest since the
problem posed leaves out of account some of the most
significant factors in the debate about resource depletion.
Here an attempt is made to outline the effects of allowing for
production and the possibility that a resource $R$ in fixed
supply may nevertheless have substitutes. The great merit of
the utilitarian models of the depletion problem incorporating
production is that they focus attention on the possibilities of
factor substitution.[27]

As above it is supposed that the utilitarian social welfare
function (SWF) takes the form

$$W = \sum_{t=0}^{T} \delta^t U(C_t).$$

But in this case utility is a function not of direct resource consumption, but of a separate consumption good. Output is assumed to be in the form of a good which can either be consumed directly or alternatively used as capital along with another factor (resource $R$) in the production of itself.[28] The production function relating total output in period $t$, $Q_t$ to inputs of capital and resources $K_t$ and $R_t$ may be written

$$Q_t = F(K_t, R_t). \qquad (3.9)$$

This output may be used either for consumption or for adding to capital, hence

$$Q_t = C_t + \Delta K, \qquad (3.10)$$

where $\quad \Delta K = K_{t+1} - K_t.$ $\qquad (3.11)$

Putting (3.10), (3.11) and (3.12) together we obtain

$$F(K_t, R_t) = C_t + K_{t+1} - K_t. \qquad (3.12)$$

As before we now impose a limitation on the stock of $R$

$$\sum_{t=0}^{T} R_t = \bar{R}.$$

The inter-temporal allocation problem can now be written formally as

$$\text{Max} \quad W^* = \sum_{t=0}^{T} \delta^t U(C_t), \, st \, F(K_t, R_t) = C_t + K_{t+1} - K_t,$$

$$\text{and} \quad \sum_{t=0}^{T} R_t = \bar{R}, \, (t = 0, 1, 2, \ldots, T).$$

This formulation is identical to that used in the 'cake-eating' example, but in addition there is now a production constraint. The Lagrangean expression can be written

$$L = \sum_{t=0}^{T} \delta^t U(C_t) + \sum_{t=0}^{T} \lambda_t [F(K_t, R_t) - C_t - K_{t+1} + K_t]$$

$$+ \lambda^* \left[ \bar{R} - \sum_{t=0}^{T} R_t \right].$$

First-order conditions may then be derived as follows:

$$\delta^t U'(C_t) = \lambda_t \tag{3.13a}$$

$$\lambda_t F_{k_t} = \lambda_{t-1} - \lambda_t \tag{3.14}$$

$$\lambda_t F_{R_t} = \lambda^*, \tag{3.15}$$

where for notational convenience $F_{k_t}$ represents the partial derivative of the function $F$ with respect to $K_t$. Assuming that second-order conditions for a maximum are fulfilled we obtain the following properties of an optimal time path.

From 3.14a

$$\lambda_t = \frac{U'(C_t)}{(1+r)^t} . \tag{3.13b}$$

From 3.16

$$\lambda^* = \frac{F_{R_t} U'(C_t)}{(1+r)^t} . \tag{3.16}$$

In equation (3.16) the numerator is the marginal product of a unit of resources $R$ multiplied by the marginal utility of consumption. This quantity might be termed the 'marginal utility product of resources', i.e. it is the extra utility which can be derived at period $t$ from using an extra unit of resources in production. The denominator is the familiar discount factor. Thus at each point in time the discounted marginal utility product of resources must be constant. The condition corresponds precisely to equation (3.8a) in the 'cake-eating' model without production. From equation (3.14),

$$\frac{\lambda_t - \lambda_{t-1}}{\lambda_t} = -F_{k_t}.$$

Substituting from equation 3.13b

$$1 - \frac{U'(C_{t-1})(1+r)^t}{U'(C_t)(1+r)^{t-1}} = -F_{k_t}$$

or $\quad \dfrac{U'(C_t) - U'(C_{t-1})}{U'(C_{t-1})} = r - F_{k_t} \left[ \dfrac{U'(C_t)}{U'(C_{t-1})} \right]. \tag{3.17}$

Once again this expression is the exact counterpart of (3.8b) in the 'cake-eating' model. The marginal utility of consumption should optimally increase at a rate equal to the pure rate of discount *minus* the second expression on the right-hand side. In order to avoid the full complexities of dynamic programming we have outlined the problem in terms of discrete time intervals. We might imagine, however, those time intervals becoming progressively shorter. It can then be shown that equation (3.17) may be written:

$$\frac{\frac{d}{dt} U'(C)}{U'(C)} = r - F_k. \tag{3.18}$$

This is the famous Ramsey Rule[29] of optimal growth. Ramsey, as we have indicated, argued that a positive pure time-preference rate was unjustifiable, in which case $r$ disappears from the equation.

Before examining in more detail the consequences of these results, one more relationship of importance can be derived.

From 3.16 we have:

$$\lambda^*(1 + r)^t = F_{R_t} U'(C_t)$$

$$\Delta[\lambda^*(1 + r)^t] = \Delta[F_{R_t} U'(C_t)]$$

or

$$\lambda^*[(1 + r)^{t+1} - (1 + r)^t] \cong F_{R_t}[\Delta U'(C_t)]$$
$$+ U'(C_t)[\Delta F_{R_t}].$$

Hence

$$\lambda^*(1 + r)^t r \cong F_{R_t}[\Delta U'(C_t)] + U'(C_t)[\Delta F_{R_t}].$$

Substituting from (3.16) and dividing throughout by $F_{R_t} U'(C_t)$,

$$r \cong \frac{\Delta[U'(C_t)]}{U'(C_t)} + \frac{\Delta[F_{R_t}]}{F_{R_t}}.$$

As before, the shorter the time intervals become the more accurate the approximation until in the limit we may write

$$r = \frac{\frac{d}{dt} U'(C)}{U'(C)} + \frac{\frac{d}{dt}(F_R)}{F_R}.$$

Substituting from equation (3.18) we then obtain

$$F_k = \frac{\dfrac{d}{dt}(F_R)}{F_R}. \tag{3.19}$$

The optimal time path will be such that the marginal product of capital will equal the *rate of increase* in the marginal product of the resource. From the discussion of the depletion-rate decision in Section 3.3 this result is not unexpected. The return from holding physical capital assets (the additional output) must be the same as the return from holding resources (their capital gain, or in this case the increase in their marginal productivity). Equation (3.19) is in effect a condition for optimal resource use viewed from the standpoint of the asset market rather than the flow of output.

Equations (3.18) and (3.19) only give us some qualitative indication of the properties of an optimal time path of consumption and depletion. As long as the marginal product of capital exceeds the pure time-preference rate it follows from (3.18) that the change in the marginal utility of consumption must be negative and, hence, given the properties of $U(C_t)$ that consumption will optimally be rising. To be more specific about the time paths of consumption, capital accumulation and resource depletion requires, of course, more detailed assumptions about the nature of the production function $F(K_t, R_t)$.

In the context of resource depletion, as distinct from the study of optimal growth in the absence of natural resource constraints, attention has naturally been focused on the question of how the model behaves as the time horizon studied gets progressively longer. In particular it is interesting to know whether consumption must eventually decline towards zero as natural resources are depleted, or whether continual growth is possible. It turns out, as might be expected, that the answer turns crucially on the degree to which natural resources and capital inputs are substitutes for one another in production. The production function most usually used to illustrate this point[30] is the constant elasticity of substitution

(CES) function, which may be written

$$Q = A[\Phi K^{-p} + (1 - \Phi)R^{-p}]^{-1/p} \qquad (3.20)$$
$$(A > 0; 0 < \Phi < 1; p > -1),$$

where the elasticity of substitution

$$\sigma = \frac{1}{1 + p}. \qquad (3.21)$$

The significance of the value of $\sigma$ can be seen more clearly by rewriting this function

$$\left(\frac{Q}{A}\right)^{-p} = [\Phi K^{-p} + (1 - \Phi)R^{-p}]$$

or $\quad \Phi K^{-p} = \left(\dfrac{Q}{A}\right)^{-p} - (1 - \Phi)R^{-p}. \qquad (3.22)$

Taking $Q$ as a fixed level of output it is possible now to investigate how capital input $K$ will have to respond to a decline in resources $R$. Clearly if $-1 < p < 0$, i.e. $\sigma > 1$, the second term on the right-hand side tends to zero as $R$ tends to zero. Even with no resource inputs, therefore, output $Q$ could be produced by using a sufficient quantity of capital. Indeed in the limit we have

$$\Phi = \left(\frac{Q}{AK}\right)^{-p}, (R \to 0)$$

or $\quad \dfrac{Q}{K} = A\Phi^{-1/p} = A\Phi^{\sigma/\sigma - 1}.$

A typical isoquant for output $\bar{Q}$ is indicated in part ($a$) of Figure 3.9. In such a case, therefore, resources are not 'essential' in the sense that output cannot be produced without them,[31] while as the input of resources declines towards zero the marginal and average product of capital tend to a positive limit and do not decline to zero. It is seen from equation (3.18) that as long as this limiting value of the marginal produce of capital exceeds the pure time-preference rate, consumption will optimally continue to grow and the

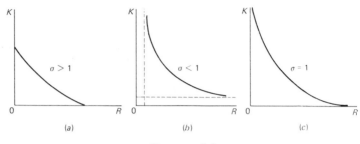

FIGURE 3.9

limited resource $R$ imposes no barrier to the onward march towards 'bliss'.[32]

Returning to expression (3.22) again it is seen that if $p > 0$, i.e. $\sigma < 1$, the second term on the right-hand side becomes larger as $R$ tends to zero. Indeed there will come a point even before $R$ reaches zero at which the right-hand side of equation (3.22) disappears or becomes negative. By increasing $K$ the left-hand side can become arbitrarily close to zero, but no increase in $K$ can make it negative. In short for each value of $Q$ there is some smallest quantum of $R$ which is required to make production possible, even with the assistance of an indefinitely great amount of capital. The typical isoquant is shown in part $(b)$ of Figure 3.9. Evidently with the substitution possibilities available in this second case the future is bleak. Even with an infinite capital stock there is a limit on the amount of output which can be produced with a given quantum of resources $R$. In these circumstances it is evident that consumption must eventually decline over time as with declining resource inputs, the marginal product of capital falls towards zero.

The final possibility is to consider the case in which $\sigma = 1$ or $p = 0$. It can be shown[33] that the Cobb–Douglas production function is a special case of a CES function with the property that $\sigma = 1$. The Cobb–Douglas function takes the form:

$$Q = A K^{\Phi} R^{1 - \Phi}.$$

In this case, although the resource $R$ is 'essential' in that production falls to zero without it, there is no upper limit to the productivity of resources as there was in the last example. Any amount of output can be produced, however small the quantity of resources used in the process as long as a sufficient quantity of capital is available. The typical isoquant is shown in part (c) of Figure 3.9. It can be demonstrated, however, that although resource productivity can rise indefinitely the marginal productivity of capital will eventually fall below any positive time-preference rate and consumption will optimally decline towards zero after this point.[34]

The utilitarian analysis of optimal depletion as outlined above concentrates only on a few of the major issues. It ignores extraction costs, assumes away technical progress and population growth and pays no attention to environmental costs or uncertainty. To incorporate such factors into an analysis would add enormously to its complexity although work along these lines is being done. Solow and Wan,[35] for example, investigate the impact of differential extraction costs on optimal depletion rates, Schulze[36] examines environmental damages and decline in resource quality, Dasgupta and Heal[37] extend their model to incorporate the probability of the perfection of a backstop technology. The basic utilitarian approach incorporating production does highlight, however, the importance of substitution possibilities between capital and natural resource inputs in the determination of the 'optimal' consumption and depletion path. This issue is central to the debate between 'optimists' and 'pessimists' on the depletion-rate problem, the former emphasising past evidence that substitution elasticities are high, and the latter arguing that whatever may have been the case in the past there is no assurance that it will remain so in the future.[38]

## (C) *The Maxi—Min Criterion*

The value judgements embodied in the additive utilitarian SWF have been severely criticised by followers of the liberal philosopher John Rawls. These criticisms are not aimed at the justification for 'pure time preference' or at the assumption of inter-personal comparability of utility. They stem instead from a different conception of justice.[39] Rawls suggests that we can gain insights into the nature of justice by envisaging individuals behind a 'veil of ignorance' drawing up an agreed contract or set of rules for the operation of a just society. Since individuals cannot know the characteristics of the state into which they will be born, or personal characteristics such as colour, sex, intelligence and so fourth, they will have no incentive to make choices in favour of narrow personal interests (since they will not know what they are). Personal interest can be pursued only by agreeing to just rules for the conduct of society. One of the most celebrated of these rules which Rawls asserts would be adopted is the maxi—min criterion for assessing the justice of inequality. No inequality is just unless it is to the advantage of the most unfortunate individual.

This principle of justice is most easily applicable to intra-temporal distribution questions,[40] and Rawls himself renounced the use of the criterion for problems of inter-generational equity in favour of alternatives. The reason for this rejection is evident — it would seem to rule out all capital accumulation as inter-temporally unjust. Such a rule might therefore be seen as a perpetuater of poverty.

Solow,[41] however, has attempted to discover the implications of adopting the maxi—min criterion where some natural resource endowment acts as a constraint. The criterion calls for equal consumption per head across generations since a rising/falling consumption path would give rise to the possibility that less/more saving by an earlier generation would improve the consumption of the least-well-off. Using a Cobb—Douglas production function Solow concludes that, with a given capital stock to start things moving[42] and a constant population, a constant consumption level is sustainable for ever provided that the share of capital in total output is greater than the share of resources.[43] This solution is contrasted with the additive utilitarian solution which, in the absence of pure time preference, will aim for rising consumption over time (see above), using the stock of resources at a slower rate but requiring a higher rate of saving.

It is in the presence of technical progress that faith in the maxi—min principle is tested most, for it implies that society will opt (justly) to run down its stock of physical capital and resources allowing technical progress to permit the achievement of constant consumption levels when a small sacrifice by an earlier generation might permit higher (possibly exponentially higher) consumption levels by future generations. Nevertheless, the Rawlsian approach can have many interpretations and that of a constant consumption path across generations is very austere. Other writers have appealed to the Rawlsian experiment of the veil of ignorance to justify alternative concepts of inter-generational equity.

## (D) *The Conservation Criterion*

The maxi—min criterion, as discussed above, was concerned with identifying a 'just' rate of savings. Inter-generational equity in the context of resource depletion, however, embraces more wide-ranging issues. The present generation may bequeath physical capital assets to the future, but at the same time it may create enormous waste disposal and environmental problems, and irrevocably destroy assets which might otherwise have yielded benefits for all future generations. Considerations such as these have led to the rise of the 'conservation ethic', a moral position the practical implications of which are often difficult to pin down. Here we merely outline briefly three interpretations which have been suggested and which are derived either specifically or indirectly from Rawls's notion of justice.

## (i) *Preservation of the Resource Base*

A principle of intergenerational justice focusing on the preservation of the 'resource base' is obviously dependent on how the term 'resource base' is defined. In the extreme version it would be regarded in physical terms so that any depletion of a non-renewable resource would fall foul of this criterion. Mankind would have to return to a state in which survival was possible from the yearly harvest of renewable resources. Economists writing in this field have tended to regard such a criterion as unnecessarily severe.[44] The important factor, it might be argued, is not the *physical* size of the resource base, but its *effective* size in terms of what can be produced from it. This has led to an alternative criterion — 'maximum sustainable yield'.

## (ii) *Maximum Sustainable Yield*

The maximum sustainable yield criterion is most easily illustrated for the use of an exhaustible but replenishable resource such as timber or fish. Justice then demands that the resource should be managed in such a way that the harvest per unit of time can be maintained indefinitely. From the economists' point of view this definition requires some added sophistications in order to take into account any natural or production cycles, and in order to ensure that the yield is defined after the costs of production and management have been subtracted.

In the case of non-renewable resources the problem immediately arises of deciding what constitutes the 'yield'. Any rate of extraction must inevitably deplete the stock in physical terms so that no constant rate of *extraction* is maintainable for ever. In so far as these non-renewable resources are inputs to the production process, however, the possibility suggests itself of defining yield in terms of the total output which is notionally producible from the remaining stock. Viewed in this way, therefore, depletion would be acceptable in so far as the productivity of the resources remaining rose sufficiently, either by technical progress or capital accumulation, to maintain the resource 'yield' in terms of potential output. Alternatively the 'yield' might be viewed simply as the total output produced per unit of time and a just depletion rate would be one which led to the maximum sustainable output. This latter conception would appear to have little in common with maxi—min justice, however, since it would appear to impose enormous burdens on earlier generations, who would be required to build up the capital stock so that later generations could enjoy maximum sustainable output levels.

A more practical version of the sustainable yield criterion as applied to non-renewable resources is suggested by Talbot Page.[45] He argues that the objective should be to maintain a price index of 'virgin material resources' at a constant level *vis-à-vis* the G.N.P. price deflator. The real price of resources considered as a group should therefore remain constant over time. This idea has some attractive features, not the least of which is the fact that it is operational. However, as Page himself recognises, it suffers from difficulties inherent in the use of index numbers. Technical innovations might render obsolete the initial 'basket' of resources upon which the index number is based, and quality changes, i.e. in this context changes in the productivity of resources, might seriously bias the index in an upward direction. A rising price index of resources is quite compatible with a constant or even declining share of income originating in the resources sector; indeed Solow's models of optimal depletion discussed in part C above (which is clearly a sustainable yield model) involves the price of resources rising continuously over time while the assumption of a Cobb–Douglas production function involves constant factor shares. Page suggests that, should the level of real expenditure on the extractive sector decline over time, the index number criterion might be relaxed a little.

### (iii) *Permanent Livability*

The concept of a maximum sustainable yield concentrates on the effect of present policies on future consumption levels. It would not in itself ensure the maintenance of 'permanent livability' through the protection of the environment. There would probably be little dispute that from a Rawlsian perspective 'permanent livability' would be likely to emerge as an agreed condition of inter-generational justice. Practical rules for achieving this end are, however, more difficult to derive. Herfindahl and Kneese,[46] to take one example, tentatively suggest the principle that no action should be taken by the present generation which 'forecloses return to the prior situation'. Present society, that is, should avoid taking decisions which entail results which are irreversible. Such a principle would forbid the present generation from (say) hunting the blue whale to extinction, but many issues are inevitably less clear-cut. The consequences of present actions are often uncertain so that differing attitudes to this uncertainty can give rise to differing prescriptions. Whether to leave the future with a legacy of radioactive waste, for example, which would contravene the principle of 'irreversibility' might no doubt be debated for a long time.

In this section we have outlined the major normative approaches to depletion-rate theory – Paretian, Utilitarian and Rawlsian. Of these, the Paretian and Utilitarian approaches are formally the most developed, perhaps because they lend themselves more readily to mathematical modelling. On the other hand the Rawlsian framework focuses attention most vividly on the broadest issues of inter-generational equity, issues which the Paretian approach ignores and which tend to be submerged in the form of the inter-temporal utility function in utilitarian analysis. The section which follows explores the *efficiency* criterion in more detail and examines how resource depletion might be affected by various types of market failure. This is done on the assumption that environmental problems, in so far as they affect the present or the future, may be adequately treated by specific measures directed to this end, measures which are explored in some detail in Chapter 5. Further, although the efficiency criterion is silent on the inter-temporal distribution of income the utilitarian and maxi—min models described above stipulate time paths for depletion which are efficient in the Paretian sense. Efficiency tends therefore to be a *necessary* condition for an optimal depletion rate even if the imposition of specific distributional judgements means that it is not *sufficient*.

### 3.5   Is Depletion Too Fast?

From the preceding discussion in Section 3.4 it will be evident that the answer to this question cannot be given in isolation from a stated objective. By emphasising *efficiency* it is not intended to convey the impression that other objectives are not important, but 'market failure' has been the subject of much comment by economists, as might be expected, and it is to the main suggested sources of this failure that we now turn.

## (A) *The Rate of Discount*

The importance of the rate of discount in public expenditure appraisal generally has resulted in an enormous literature on this topic.[47] From Section 3.3 it is seen that, if anything, the question of the discount rate is even more important in the field of resource depletion. Here we merely outline the major arguments adduced for thinking that market rates of discount will not represent the 'efficient' rate.

### (a) *Social v. Private Risk*

It can be argued that rates of discount used in the market are likely to be too high if there is inadequate provision for 'risk pooling'. Suppose two individuals are contemplating risky projects. Both projects have a positive present value when expected returns are discounted at a 'riskless' rate of interest. Neither project is undertaken, however, because both individuals discount at a higher rate to take account of the risks which they face. Suppose now that the pay-off on each project is identical in size and depends simply on whether or not a particular state of the world occurs at a given future point in time. Suppose further that the probability of this state is known and that where one project succeeds the other must fail. Clearly, although the projects considered singly are risky, the projects considered together will have a pay-off that is certain. If financial markets are operating efficiently this should not pose serious problems. The two individuals might simply decide to own a share in each project, thereby eliminating the risk. Equivalently they might each insure against the possibility of the 'wrong' state of the world occurring.

But markets it is argued do not operate so smoothly. Not only do brokerage and other administrative costs inhibit the growth of markets in 'state contingent claims' (i.e. claims to a specified return contingent upon the occurrence of a particular state of the world), but problems of 'moral hazard' make the policing of such markets costly.[48] It is therefore to be expected that the private market rate of discount will exceed the efficient rate as individuals allow for risks which, from the point of view of society as a whole, may merely represent transfers from one person to another.

The absence of well-developed futures markets is perhaps the most powerful reason to expect 'market failure' in the allocation of natural resources. Individual owners of resource deposits face the possibilities that technical change will result in substitutes, that new deposits will be discovered adversely affecting the value of their asset, and that governments may take over the resource with inadequate compensation. Some of these risks are negligible from a social point of view. True, it might be argued that if a 'backstop technology' is invented when some unused deposits of fossil fuel remain unexploited the population will have lost the benefit that they might have experienced by using these deposits more quickly. But, as Ivor Pearce argues, this risk is negligible in comparison with the risk to the owner of the resource who finds the value of his asset wiped out. On the other hand if the resource owner had a sufficiently diversified portfolio of assets, as was suggested earlier, the loss of wealth to any individual might still be small and perhaps compensated for in other areas. Assuming insufficient diversification it would appear that risk-averse resource owners will discount the future at too high a rate and that the expected trend in prices must increase at this rate. The implication is clearly that resources will be depleted too quickly in these circumstances.[49]

### (b)  *The Isolation Paradox*

The isolation paradox, propounded by Sen[50] (1961) and Marglin[51] (1963), arises from a form of consumption externality. It is suggested that individuals living now will gain satisfaction from contemplating the wellbeing of future generations. Since an individual act of saving (and investment) will benefit the future it confers also an external benefit on contemporaries. Saving will therefore be sub-optimal in the sense that all would *agree* to save more if the institutional arrangements were available, but no *one* individual has an incentive to increase his saving rate (hence the paradox). Each individual will be willing to save more providing all other members of the community join him. It is then inferred that, since investment is sub-optimal, the rate of interest in the market must be inefficiently high.

## (c) *Taxation*

Taxation of income from capital will lead to an inefficient divergence between the post-tax return and the pre-tax return on investment as we have already indicated in Section 3.2. Again the conclusion is drawn that market interest rates are too high, investment sub-optimal and depletion consequently too fast.

However, companies operating in natural resource industries are typically treated rather differently from those operating in other spheres and this complicates matters considerably. Full discussion of tax policy with reference to energy resources must be deferred to Chapter 6. Here we observe that in the United States tax measures such as percentage depletion, the importance of capital gains tax rates to mining companies, and the 'expensing' of exploration and other costs, have probably resulted in a more rapid depletion of energy resources than would otherwise have occurred. In the United Kingdom coal has been historically the most important indigenous fossil fuel. Since nationalisation in 1946 output levels have declined enormously as a result of competition from alternative fuels so that depletion has hardly been an important issue. In recent years, however, the discovery of oil and gas deposits in the North Sea has put the U.K. Government in the unfamiliar position of having to consider the implications of their tax policies for development and depletion.

## (B) *Property Rights*

(i) *Effects on production*   Drilling for oil and gas is particularly vulnerable to the problem of establishing a clear property right in the resource being exploited. Where property rights are not clearly defined excessive use of the resource is predictable (air or water pollution and over-fishing are other examples of the same problem) and competitive drilling in a single reservoir of oil presents a classic case of an externality.

It is clearly not in the interests of a single well-owner to conserve output in the manner described in Section 3.3 if by delaying production it is for ever lost to some other local competitor. The result will be a scramble to drill a large number of wells and deplete the source as quickly as possible. Recognition of this problem has led in the United States to complex regulations limiting well-spacing and production, and permitting 'unitization', i.e. agreements between owners to run oilfields as single units.[52]

(ii) *Effects on Exploration*    Similar problems to those above arise with respect to the exploration decision. Two possibilities are usually suggested, one leading to too low and the other to too high a level of exploration.

(*a*) It can be argued that since knowledge is itself a good over which it is exceptionally difficult to establish a property right, geological investigation will not be carried to a socially optimal level. A survey of a relatively unknown area may provide information of considerable value not only to those undertaking the survey but to other enterprises with interests in natural resources.

(*b*) A consideration militating in the reverse direction is the system used in the United States for establishing the right to work a particular deposit — the 'location system'. At least for certain minerals this entails actually discovering a source and then claiming the mineral rights. In a competitive environment the tendency will therefore be to explore for new deposits as soon as the expected present value[53] of the find is positive, even though there may be no intention of exploiting it for several years. It is evident that this premature exploration will entirely remove the 'user cost' or 'royalty' element which would otherwise have arisen. The efficient policy would involve the sale of mineral rights by competitive bidding thus enabling the Government to receive the royalty (or at least its expected value) and permitting the company with the rights to delay exploration until the appropriate time. The competitive bidding system has been used in the United States for the Federal offshore drilling programme. In the United Kingdom, on the other hand, the licensing system in the North Sea has been somewhat different, with an emphasis, especially in the early years, on specifically encouraging exploration.[54]

## (C) *Stability*

The absence of perfectly functioning futures markets results, as has been noted, in individual operators having to rely on their 'expectations' about future prices rather than definite information from forward contracts. The question then arises as to how close to the efficient path the economy will follow relying on these expectations. Suppose, for example, that at each period expected prices are rising at the appropriate rate to keep capital markets in equilibrium and that these expectations are always confirmed in the following period. If the initial price of the resource was too high or too low the economy might continue along such a path for a very long time before any mechanism revealed the 'mistake'.[55]

Alternatively less placid circumstances might be envisaged in which expectations of future price increases became higher or lower than the rate of interest. If expected prices are rising at too slow a rate the analysis of Section 3.3 indicates that output will be increased in the present period since unworked deposits are a bad investment. Such an increase in production will depress the current price a factor in itself which might tend to restore equilibrium. However, everything will depend on the response of expectations to this fall in price. If expectations are rather elastic and expected future prices fall along with current prices the market will still be out of equilibrium and a cumulative downward spiral of spot prices could continue. This somewhat pessimistic analysis suggests that markets in natural resources might be prone to explosive price movements either up or down as a result of fairly small changes in expectations. However, the precise properties of any model obviously depend crucially on how expectations are formed. Instability in the case above occurred because no one was willing to take a long view and stick to a particular opinion in the face of current fluctuations. Where there exist individuals whose expectations are governed by longer-term considerations than changes in the current price, transactions in asset markets, i.e. the market for deposits, will act as a stabilising force, as Solow points out.[56] In this case a small change in the expected rate of price increase might result in an adjustment in asset values sufficient to make the new expected, price regime consistent with the 'fundamental principle'.

An alternative view put forward by Kay and Mirrlees[57] suggests that price predictions are unlikely to be too difficult as long as the reserves of a particular resource are sufficiently large. Again in Section 3.3 it was noted that larger reserves will increase the time to exhaustion and reduce the royalty. Kay and Mirrlees point out that, with a real interest rate of 5 per cent, reserves of 250 times current consumption and an assumed price elasticity of unity, the royalty would remain below 1 per cent of marginal extraction costs for 130 years and for all practical purposes might as well be ignored. The argument, however, is very sensitive to the interest rate used. A real rate of interest of 2 per cent, for example, would be sufficient to eliminate the 130-year period prior to the emergence of a 'user cost' of any significant size.

## (D) *Monopoly*

Markets in energy resources bear little relation to the perfectly competitive model of textbook theory. Oil production, for example, is dominated by a cartel of producing countries — OPEC — and the structure of the refining industry is oligopolistic. Predictions about the behaviour of cartels and oligopolies are notoriously difficult to make, but assuming a fairly 'tight' organisation and a joint profit-maximising objective it would be expected that output would be lower at any given time compared with a competitive environment. Kay and Mirrlees rely heavily on this argument when they conclude that 'resource depletion often takes place too slowly because of monopoly power exerted by resource owners'.[58]

## 3.6 Conclusions

One thing at least can be inferred from the preceding discussion: there is most unlikely to be agreement about how fast resources ought to be depleted. Ideas on this issue will vary with value judgements concerning the weight to be given to the wellbeing of future generations and with views about the nature of the economic system itself — whether substitution possibilities are extensive and growing or whether choices are becoming increasingly limited and technical progress less likely.

Even accepting the inter-generational distribution of income as given, and concentrating in true Paretian fashion on inter-temporal efficiency as an objective, we have noted many reasons to expect markets to function inefficiently. On the other hand no attempt has been made to investigate alternative institutional arrangements. It would be easy to succumb to the 'nirvana fallacy' at this point and recommend extensive government intervention to correct the inadequacies of the market. We shall argue that there are important areas for government intervention, but it should always be remembered that governments too suffer from most of the limitations of the market-place.[59] If future generations do not bid directly in existing markets, neither do they vote in present elections, and indirect influence may well be stronger in the former than the latter case. If market operators are risk-averse and deplete resources too quickly governments will have their own special reasons to prefer cheaper to more expensive energy, especially in representative democracies. It is also as well to note that most abstract theorising views energy as a world problem, whereas actual policies are implemented by national governments concerned with something less than world welfare.

Ultimately therefore the depletion-rate issue becomes a complex problem in political economy in which institutional factors as well as abstract models play their part. The type of analysis described in this chapter is required, however, to clarify the logical implications of given objectives and to highlight the role played by particular factors such as substitution possibilities.

# 4. ENERGY PRICING

## 4.1 Introduction

The choice of an energy pricing policy is concerned with the answering of a number of interrelated questions such as: On what basis should energy prices be set? What should be the relationship between the prices of the various forms of energy? At what rate should a fossil fuel be depleted? To what extent should energy prices be used to achieve equity objectives? Should energy prices reflect associated environmental costs? It is thus clear that the subject-matter of this chapter in inextricably linked with that of the preceding and following chapters.

Questions such as these can only be answered in the context of a specific choice criterion which allows the ranking on a scale of better and worse of the various consequences associated with different pricing policies. In addition they càn only be answered in a specific institutional setting. This will define the circumstances within which energy pricing decisions have to be taken, such as whether the industries are in the public sector or in the private sector but subject to regulatory control, and any constraints which may limit the freedom of choice.

The determination of 'optimal' energy pricing policies is complex. These policies have to be formulated for multi-product industries (e.g. different grades of coal, high- and low-voltage electricity, different types of oil) which supply both final product and intermediate goods. While some of these outputs are storable at economic cost levels (coal, gas and oil), electricity tends to be non-storable at such levels, and pricing policies must allow for whether the product is storable or non-storable. Energy prices must also allow for the different possible uses of primary energy resources. Thus natural gas can be used in either the premium or non-premium

heating markets or, for example, as an input into the manu-
facture of fertilisers.

This chapter is in eleven sections, as follows. Some aspects
of the interdependency of pricing and investment decisions
are considered in Section 4.2. The specification of objectives
and constraints is considered in Section 4.3. Marginal cost
pricing is considered in Section 4.4, while Section 4.5 considers
peak-load pricing. The analysis of Section 4.5 is modified in
Section 4.6 by allowing for feasible storage and system
effects. Some aspects of the practical relevance of the analysis
of the preceding sections is considered in Section 4.7, and
some problems relating to the measurement of marginal costs
are considered in Section 4.8. Sections 4.9 and 4.10 are
respectively concerned with the modification of energy prices
to achieve financial targets and income distribution objectives.
Finally, some conclusions are presented in Section 4.11.

## 4.2   Interdependence of Pricing and Investment Decisions

Although this chapter is concerned with pricing problems it is
important to remember that pricing and investment decisions
are interrelated. This is illustrated in Figure 4.1. Providing
that the own price elasticity of demand is not equal to zero
different prices (for example, relating them to either average
or marginal costs when these are unequal) will lead to different demand forecasts, different optimal investment programmes, different cost estimates and hence prices. In
principle it is possible to use an iterative process to ensure

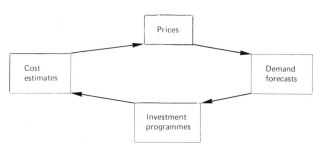

FIGURE 4.1

consistency between the values of prices, demand forecasts, etc. But in practice there are frequently relatively long time lags in the energy industries between, for example, changes in prices and consequent changes in demands because of the time required for consumers to adjust their stocks of energy-using appliances. In that case it is reasonable to calculate an 'optimum' (for given objectives) set of prices for an assumed investment programme.

While this may be a satisfactory basis for the setting of prices and the choice of investment programmes, the factor of interdependency is of crucial importance. The relating of prices to different bases will have a number of important effects. Thus the use of different bases will lead to different allocations of resources, to different distributions of incomes (since all pricing policies have distributional implications), to different financial outcomes for the industry, to different effects on a country's balance of payments and on the level and structure of its employment, etc. It thus follows that before decisions can be taken on what constitute appropriate energy pricing policies the objectives and any relevant constraints for the energy utilities must be specified. Before we consider this specification it is worth considering whether energy prices have allocative significance.

To pose this question is basically to ask whether the evidence on energy price elasticities shows them to be different from zero. Neoclassical demand theory predicts that an increase in the general price level of energy relative to other prices would reduce its consumption, and similarly that a relative increase in the price of one fuel would tend to reduce its share of the energy market. Time trend data for the United Kingdom for the 1960s and 1970s shows that in domestic and industrial markets there was a substantial shift away from coal to gas and oil, and that this shift was accompanied by a substantial change in the relative prices of these fuels against coal. The use of this data to calculate energy price elasticities presents a number of difficult problems.[1] These include that for the United Kingdom until the early 1970s changes in energy consumption were dominated by changes in income since, with the exception of gas prices, energy prices exhibited only smooth fluctuations about smooth time trends. 'It is thus impossible to distinguish meaningfully the effects of any long-term change in prices from the effects of income. . . . '[2] In addition, the elasticities are unlikely to have remained constant over time since for the United Kingdom there have been marked changes in the availabilities of different fuels, especially the availability of off-peak electricity and natural gas in the 1960s. These changes in availabilities will have affected the price elasticities of other fuels.

These and other problems have been compounded in the 1970s by the large increase in relative energy prices in and since 1973. These price increases cast considerable doubt on the applicability to the post-1973 period of energy price elasticities calculated for periods before 1973. While it is acknowledged that the numerical estimates of elasticities calculated for periods pre-1973 are almost certainly not applicable to periods post-1973 this evidence is relevant to the question posed above.[3]

For illustrative purposes Table 4.1 presents some estimates of own and cross price elasticities for the United States. The estimates were calculated using cross-section data for 49 states for the period 1968–72 and are averages for the different states. These estimates have the expected signs and they indicate that energy prices do have allocative significance. They are broadly consistent with those of a number of other authors. For example, using time-series data for 1961–9 for 48 states R. Halvorsen[4] estimated the long-run elasticity for residential electricity to be between −1.0 and −1.21. Rather lower estimates were produced by J. M. Griffin.[5] Using time-series data for 1951–71 he estimated the long-run residential elasticity for the United States to be −0.52.

The qualitative importance of these estimates is supported by data for a number of other countries. Thus using time-series data for 1954–72 A. S. Deaton[6] estimated the residential short-run elasticity for electricity in the United Kingdom to be between −0.30 and −0.96 for 1970, and for residential gas to be between −2.64 and −2.90. For New South Wales and the Australian Capital Territory R. G. Hawkins,[7] using cross-section data, estimated the long-run price elasticity for residential electricity demand to be −0.55 in 1971.

*The Economics of Energy*

TABLE 4.1

*Energy Price Elasticities for the United States, 1968–72*

| Residential and commercial | | Price Gas | Price Oil | Price Electricity | Price Coal |
|---|---|---|---|---|---|
| Gas consumption | SR | −0.15 | 0.01 | 0.01 | n.a. |
| | LR | −1.01 | 0.05 | 0.17 | n.a. |
| Oil consumption | SR | 0.04 | −0.18 | 0.01 | n.a. |
| | LR | 0.19 | −1.12 | 0.16 | n.a. |
| Electricity consumption | SR | 0.05 | 0.01 | −0.19 | n.a. |
| | LR | 0.17 | 0.05 | −1.00 | n.a. |
| *Industrial* | | | | | |
| Gas consumption | SR | −0.07 | 0.01 | 0.03 | 0.01 |
| | LR | −0.81 | 0.14 | 0.34 | 0.15 |
| Oil consumption | SR | 0.06 | −0.11 | 0.03 | 0.01 |
| | LR | 0.75 | −1.32 | 0.34 | 0.14 |
| Electricity consumption | SR | 0.06 | 0.01 | −0.11 | 0.01 |
| | LR | 0.73 | 0.13 | −1.28 | 0.14 |
| Coal consumption | SR | 0.06 | 0.01 | 0.03 | −0.10 |
| | LR | 0.75 | 0.14 | 0.33 | −1.14 |

SR = short-run (one-year) elasticity.
LR = long-run elasticity.

SOURCE: P. J. Joskow and M. L. Baughman, 'The Future of the U.S. Nuclear Energy Industry', *Bell Journal of Economics* (Spring 1976).

This and other evidence on elasticities suggests that relative energy prices do have allocative significance and that they play an important role in determining both the pattern and level of energy consumption.[8] Thus if in some sense energy prices are too low energy consumption will be too high.

## 4.3  Objectives and Constraints

The energy utilities are typically multi-product enterprises and their pricing policies are concerned with both the level and structure of their tariffs. We have already observed that different pricing policies for energy will have different effects on, among other things, the allocation of resources, the distribution of income and on the financial performance of the utility. Traditionally in most countries energy pricing policies have been judged solely in terms of their effect on the financial performance of the enterprise. The so-called 'energy crisis', with its emphasis on the relative scarcity of many energy resources, has brought to the fore the need to consider the allocative effects of different pricing policies.

For most of this chapter we shall assume that the relevant objective is economic efficiency in the allocation of resources. This is defined in the Paretian sense, so that an allocation of resources will be said to be efficient if it cannot be reallocated to make one consumer better-off without making another consumer worse-off. In Sections 4.9 and 4.10 we consider the introduction of financial and income distribution objectives. Note that the efficient allocation of resources between the different energy industries in a country will require that they all pursue a common set of objectives.[9]

Energy prices must be formulated given certain constraints. These include the usual ones of available resources, the state of technology, etc. Another, and important, constraint is that price structures must be sufficiently simple for consumers to be able to understand, and hence react to, them. The role of prices as signalling devices would be lost if price structures were so complicated that consumers could not understand them. In addition the price structures and levels must be socially acceptable and equitable.

## 4.4 Marginal Cost Pricing

The recommendation that energy prices should be related to their marginal social costs given the efficiency objective can be derived from economic models which are formulated in either general or partial equilibrium terms. Since the analysis of this book is couched in partial equilibrium terms we shall use that framework for the derivation of the marginal cost pricing rule. The following analysis assumes that all the Pareto-optimal conditions are satisfied elsewhere in the economy, and thus we retain the assumptions made in Section 3.2. The optimal pricing rules for any particular energy utility can then be derived as follows.

For any energy utility the social welfare function (SWF) can be written as the maximisation of

$$W = TR + S - TC,$$

where　　$W$ = net social welfare

$TR$ = total revenue

$S$ = consumers' surplus

$TC$ = total costs.

Thus welfare is equal to social benefits minus social costs, and on the assumptions made to the maximisation of the sum of consumers' and producers' surpluses. It is thus concerned with questions of economic efficiency rather than of equity. If the supply of factors of production to the utility is perfectly elastic there will be no intra-marginal rents. Total costs will then be total money costs. To maximise this function differentiate it with respect to output and set the result equal to zero:

$$\frac{\partial W}{\partial Q} = \frac{d}{dQ}(TR + S) - \frac{d}{dQ}(TC) = 0.$$

Now $TR + S$ is equal to the area under the demand curve. Let $P(Q)$ be the demand curve, so that $TR + S = \int P(Q)dQ$. Differentiating this expression with respect to $Q$:

$$\frac{d}{dQ}(TR + S) = \frac{d}{dQ}\int P(Q)dQ$$

$$= P(Q).$$

Since $P(Q)$ is price and $d/dQ(TC)$ is marginal cost we have the result that $P - MC = 0$. Differentiating with respect to $Q$ a second time shows that when price equals marginal (social) cost welfare is maximised.

It is clear that this derivation depends on the making of a number of very restrictive and unreal assumptions about the real world. If allowance is made for the fact that not all the Pareto-optimal conditions are satisfied throughout the economy (because of the existence of monopolies, non-lump-sum taxes, external effects, etc.) then the previous first-best optimisation model must be replaced by a second-best optimisation model. In that case it turns out that usually (but not always[10]) the simple prescription that the efficiency objective can be achieved (or welfare maximised) if all energy prices are set equal to their marginal costs has to be replaced by a number of sometimes very complicated pricing rules.[11] However, the purpose of most of these rules is to determine how energy prices should deviate from their marginal costs when, for example, some other prices in the economy are not set equal to their marginal costs. This suggests that a workable approach to the determination of the set of energy prices given the efficiency objective is, first, to calculate the relevant set of marginal costs and, second, to consider how prices based on these costs may need to be adjusted given various distortions existing within the economy and for the attainment of other objectives. This is the approach which is adopted in this chapter.

Ideally marginal cost is a measure of the value to society of the extra resources required to produce another unit of output in a particular time period. It is a money measure of the value of the output sacrificed elsewhere by producing another unit of the good. As we have seen in Chapter 3, for a depletable resource it will be given by the sum of two separate costs: (i) the marginal extraction cost and (ii) the present value of the net earnings forgone per unit of the resource by not leaving it in the ground. In terms of the efficiency objective the general presumption is that if the price which a consumer is willing to pay for another unit exceeds the value of the extra resources required to make it, then the allocation of resources will be improved if that unit is produced, and vice versa. The consumer's purchase decision is then based upon a consideration of relative resource costs. Energy prices based on marginal costs inform consumers of which forms of energy can be used to satisfy their demands at relatively low resource costs and which can only do this at relatively high resource costs.

Notice that energy prices based on marginal costs are concerned with the resource commitments to meet a change in demand, are forward-looking and are related to forecasts. This is in contrast to prices which are based on average accounting costs, which are backward-looking and generally related to historic measurements.

In both theory and practice the relating of prices to marginal costs poses a number of difficult problems. These include problems associated with the measurement of marginal costs (Section 4.8), and the choice of price structure when to have prices at all times equal to marginal costs would involve structures which were too difficult for consumers to understand and too expensive to administer (Section 4.7). At this point we shall content ourselves with the consideration of a simple question, namely whether marginal cost is unique, the analysis of which provides some useful inputs into later sections of this chapter.

*Is Marginal Cost Unique?*

The answer to this question is clearly 'no'. Marginal cost is a multi-dimensional concept; it typically varies with the period over which it is measured (short or long run); whether a demand increment which is being costed is permanent or temporary; with the length of forewarning which an enterprise has of a demand change since this affects its ability to undertake the optimal investment programme, etc. In terms of the analysis of the following two sections it will be useful to illustrate the non-uniqueness of marginal cost with respect to whether a utility has either an excess or a shortage of inherited capacity.

In Figure 4.2 we consider a utility which has an inherited capacity of $0q_1$ from plants with a homogeneous production technology each of which exhibits a rigid output limit and constant returns to scale. Technology and relative prices are assumed to be unchanging and there is perfect knowledge. The unit running cost (short-run marginal cost (SRMC)) of this inherited capacity is constant at $0b$ up to the output limit, when it in effect becomes infinite. (For the present

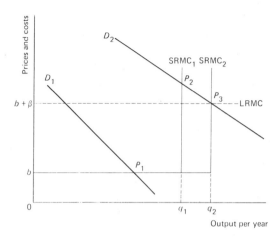

FIGURE 4.2

ignore the line designated long-run marginal cost (LRMC)
($b + \beta$) and capacity $0q_2$.)

There are two cases to be considered. First, that when the
utility has excess capacity and the demand curve is that
shown as $D_1 D_1$. The optimal price is then equal to $P_1 = 0b$.
All the consumers are 'free riders' and make no contribution
towards the financial charges of the inherited capacity $0q_1$.
In this case the optimal price is set equal to short-run marginal
cost measured with reference to resource opportunity costs.
The second case is that when the utility has a shortage of
capacity and the demand curve is that shown as $D_2 D_2$. In
this case the price that maximises the sum of consumer and
producer surpluses is $P_2$, and it rations available output
among potential consumers on the basis of willingness to pay.
Clearly price $P_2$ is not equal to the resource marginal
opportunity cost of supplying another unit, since this is
undefined at output $0q_1$. However, there is an alternative
and sympathetic concept of opportunity cost which can be
used and which permits us to say that $P_2$ is also equal to
short-run marginal cost. This concept is that of marginal user
opportunity cost,[12] and it measures the willingness to pay
for the $q_1 + 1$ unit by the marginally excluded consumer.

Case 2 clearly raises the question of whether the inherited capacity is optimal. Assume that capacity is perfectly divisible and that the capital cost of one unit of capacity is $K, with a unit running cost of $0b$. This capacity will have a life which extends over many years, but in Figure 4.2 output is measured on an annual basis. It is thus necessary to express $K$ on an annual equivalent basis so that it can be added to the unit running cost to give a measure of long-run marginal cost. The use of the unit of capacity for one year involves two costs, an interest cost and a depreciation cost. If we assume unchanging technology and prices then the first-year equivalent cost of the unit of capacity can be written:

$$\beta = rK + A,$$

where $r$ is the annual rate of interest, and $rK$ is thus the first-year interest cost, and $A$ is the first-year depreciation, which is measured by finding the net present value of the unit of capacity at the beginning of the first year and again at the beginning of the second year and taking the difference (i.e. $A = NPV_1 - NPV_2$).

For the assumed social welfare function capacity will be optimal when $P = b + \beta$ (the willingness to pay for a unit of output equals the long-run marginal cost of supplying that unit) and investment should occur if $P > b + \beta$, and disinvestment in the opposite case.

In Figure 4.2 at output $0q_1 P_2 > b + \beta$ and thus the capacity is sub-optimal and investment should occur. The optimal capacity is $0q_2$ and the optimal price $P_3$. Notice that if short-run marginal cost is defined in the marginal-user opportunity-cost sense then $P_3$ is simultaneously equal to short- and long-run marginal cost. This result is not peculiar to the specific assumptions which have been made. In the absence of indivisibilities it is a common result.

The preceding analysis has been based upon the assumption that there is a single demand function per demand cycle, which in this case lasts for one year. This assumption is unrealistic for many of the energy utilities and it is necessary to extend the analysis to allow for the division of the demand cycle into a number of sub-periods.

## 4.5    Peak-load Pricing

The demands for the products of the various energy industries typically vary in a systematic way over time. Thus in temperate countries such as the United Kingdom the demand for coal is higher in the winter than in the summer months. The demand for electricity and gas typically varies over the day, week and year (seasonal factor). Taken by itself this feature of variable demands may not be very important, since if the output can be stored at economic cost levels the variations in demand can be met by adjusting inventory levels – but see Section 4.6, below, on feasible storage. If, however, storage is not possible at economic cost levels then these variations in demand may be important. In these circumstances if the output capacity of the industry cannot be continually adjusted to keep price continually equal to short-run marginal cost the so-called peak-load problem is encountered.[13]

A number of separate cases can usefully be distinguished in the analysis of this problem. One major division is between what are termed the *firm-peak case* and the *shifting-peak case.* The essential difference between these two cases is simply that a change from a uniform to a time-differentiated tariff in the former case does not change the period responsible for the peak demand, whereas in the latter case for some set of prices the previous off-peak demand may become the peak demand. It has been found that the shifting-peak case is particularly pertinent to electricity supply (see Figure 4.6).

The peak-load problem is essentially concerned with the allocation of joint costs and arises from a problem of indivisibility. A unit of capacity, such as a power station, which is provided to meet the peak-period demands will also be available to meet the off-peak demands. This poses a number of problems, namely:

(i)   what prices should be charged for energy?

(ii)  how many separate pricing sub-periods should be distinguished in a given demand cycle (the structure of prices)?

(iii) what is the optimal capacity?

Questions (i) and (iii) will now be considered for both the

firm- and shifting-peak cases. We continue to assume the satisfaction of all the first-best optimum conditions. Consideration of question (ii) is deferred to Section 4.7.

### Firm-peak Case

This can be analysed using Figure 4.2 (p. 83). We retain all the previous assumptions with the exception that it is now assumed that the demand cycle lasts for 24 hours with two equal duration sub-periods 1 and 2, off-peak and peak respectively with demand curves $D_1$ and $D_2$ (the analysis can easily be extended to any number of sub-periods). The sub-period demands are independent of each other[14] (the cross-price-elasticity equals zero). A single price is to be charged in each sub-period. Previously $\beta$ represented the annual equivalent cost of a unit of capacity. Since the demand cycle is now assumed to last for 24 hours we let $\beta'$ represent the *daily* equivalent cost of a unit of capacity ($\beta' = \beta/365$). However, if output and unit operating costs are to be measured in terms of the duration of each sub-period (12 hours), then capacity costs must also relate to this time period. The required figure is given by dividing $\beta'$ by 2 to give $\beta''$. The daily cost of the utility is then:

$$C_1 = bq_1 + \beta'' q_1^0 \qquad 0 \leqslant q_1 \leqslant q_1^0$$

$$C_2 = bq_2 + \beta'' q_2^0 \qquad 0 \leqslant q_2 \leqslant q_2^0$$

$$C = C_1 + C_2 = b(q_1 + q_2) + 2\beta'' q^0 = b(q_1 + q_2) + \beta' q^0,$$

where    $C_1$ is total cost in period 1

          $C_2$ is total cost in period 2

          $q_1^0$ is the capacity limit in period 1

          $q_2^0$ is the capacity limit in period 2

and    $q_1^0 = q_2^0$ since the installed capacity is the same in both periods.

In Figure 4.2 inherited capacity is $0q_1$. The capital charges associated with this capacity are irrelevant to the determination of the optimal set of prices, since they do not involve any resource costs, bygones are bygones. The optimal prices are $P_1$ and $P_2$ for the off-peak and peak periods respectively. While the off-peak price is equal to short-run marginal cost in the resource opportunity cost sense the peak-period price is equal to it in the user opportunity cost sense since it is set at the level required to ration available output among potential consumers.

For the given assumptions and a single demand function per demand cycle we have previously seen that capacity is optimal when $P = \text{SRMC} = \text{LRMC}$ and that investment is warranted if $P = \text{SRMC} > \text{LRMC}$. The extension of these results to the firm peak pricing case is straightforward.

Capacity extensions are worth while if the incremental benefits exceed the incremental costs. Investment is thus warranted if the sum of the two period prices exceeds the sum of the capacity's marginal operating costs in the two sub-periods and its marginal investment cost over the demand cycle. That is if $P_1 + P_2 > 2b + 2\beta''$, and capacity is optimal when this inequality becomes an equality. In Figure 4.2 optimal capacity is $0q_2$.

Notice that in this case the utility just breaks even (since average and marginal costs are equal by assumption) and that all the capacity costs are borne by the peak-period consumers. Although this latter result has been derived assuming a homogeneous production technology it can be shown that it continues to hold with heterogeneous production technology providing that the optimal running times of the different types of capacity are equal to the specified demand sub-periods.[15]

The preceding analysis can be simply extended to allow for uncertainty when this is catered for by the provision of a reserve plant margin (as is usually the case in electricity supply). Let the size of the margin as a proportion be $\alpha = q_1 q_2 / 0q_2$, where $q_2$ is the inherited capacity and $q_1$ is the available capacity net of the reserve plant margin. Capacity will be optimal when $P_1 + P_2 = 2b + 2\beta''/1 - \alpha$. In the firm-peak case with $P_1 = b$ then $P_2 = b + 2\beta''/1 - \alpha$. That is, peak-period consumers bear all the costs associated with the provision of the reserve plant margin.[16]

### Shifting-peak Case

If the foregoing pricing solution was applied to the conditions illustrated in Figure 4.3 the lower price would be charged to the demands which made the peak demand on capacity! The pricing solution of the firm-peak case would in these conditions make the previous off-peak demand become the peak demand.

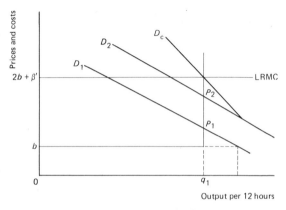

FIGURE 4.3

In these circumstances the optimal prices and capacity can be determined as follows. Since the capital cost of a unit increase in capacity is joint to the two periods in which it will be used, investment can be justified by either the demand in any single period or by the sum of the separate period demands. The demands of the various periods are thus complementary. The factor of jointness in supply means that the total demand for capacity can be obtained by summing the individual demand curves vertically, which gives the demand curve $Dc$. The optimal capacity can then be determined by equating the sum of the demand prices $(P_1 + P_2)$ to the cost of providing an increment in output over the entire demand cycle, $2b + \beta'$. This cost exceeds that of providing for an increase in either period alone, which is $b + \beta'$.

In Figure 4.3 the optimal capacity is $0q_1$ since $P_1 + P_2 =$ $2b + \beta'$ (where the latter equals the long-run marginal cost of providing for an increase in output in both periods). Rearranging this equality, capacity is optimal when $(P_1 - b) + (P_2 - b)$ $= \beta'$, and investment is warranted when $(P_1 - b) + (P_2 - b)$ $> \beta'$.

Comparing this solution with that of the firm-peak case it can be seen that both prices now exceed the unit operating cost $(b)$ and that capacity is fully utilised in both periods.

### 4.6 Feasible Storage with a Heterogeneous Production Technology

Some forms of energy for which the demands fluctuate systematically over time can be stored at economic cost levels, for example coal and gas. We will now consider what effect feasible storage has on the optimal peak and off-peak prices.[17]

We continue to assume two 12-hour independent demand functions, certainly, that all capacity exhibits rigid output limits and that the unit operating cost of each item of capacity is constant up to the capacity limit. However, we now assume a heterogeneous production technology; that is, the industry has an inherited stock of rigid plants of different vintages and technical types which are operated as an integrated supply system. The marginal running costs of newer plants will be assumed to be lower than those of older plants, and to minimise the cost of meeting demand in any sub-period they will be operated in merit order. This means that in any period plants will be brought into operation in inverse order of their short-run marginal costs, low-cost plants being run for longer than high-cost plants. Although the plants are assumed to exhibit rigid output limits it is assumed that relative to the size of the system there are so many plants that all cost curves can be drawn without any discontinuities. In such a system if the pattern and annual rate of output is held constant the introduction of a new unit of capacity with a lower marginal operating cost than any inherited capacity will lead to a saving in system running costs. This is because low-cost output from the new plant will replace high-cost output from inherited plant. This potential saving in system running costs should be credited to the new capacity in the investment appraisal calculations. We will now show that in an integrated supply system providing price in each sub-period is set equal to the system short-run marginal cost in that sub-period (the marginal operating cost of the highest cost plant which is operating in merit order in that sub-period), then when capacity is optimal the sum of these prices will equal long-run marginal costs.[18]

Assume two equal-duration sub-periods with associated demand functions $D_1$ and $D_2$ (off-peak and peak) and consider the merit of constructing a new plant with an output of $Q$ units per 12 hours. Let the capital cost of this plant be $K$ over the 24-hour demand cycle and its constant unit running cost be $b'$. Assume that prices are always set equal to the short-run marginal cost on the system ($m$), so that $p_1^1 = m_1^1$ and $p_1^2 = m_1^2$, where the subscripts denote the demand sub-period and the superscripts refer to the periods before and after the construction of the new capacity. Finally, assume that each demand curve and the system short-run marginal cost curve are linear.

Let the increment in peak-period output following the construction of the new plant be $V$ and for the off-peak period be $W$. The consumers' willingness to pay for the extra output will thus be:

$$\tfrac{1}{2}(p_1^1 + p_1^2)W + \tfrac{1}{2}(p_2^1 + p_2^2)V.$$

The cost of operating the new capacity in each sub-period will be $b'Q$, but this does not represent the increase in system costs since the capacity will earn running cost savings. Allowing for this the change in the off-peak period operating costs will be $b'Q - \tfrac{1}{2}(m_1^1 + m_1^2)(Q - W)$ and in the peak period $b'Q - \tfrac{1}{2}(m_2^1 + m_2^2)(Q - V)$.

The new capacity is worth constructing if the associated increase in benefits exceeds the associated increase in costs,

$$\tfrac{1}{2}(p_1^1 + p_1^2)W + \tfrac{1}{2}(p_2^1 + p_2^2)V > K + 2b'Q - \tfrac{1}{2}(m_1^1 + m_1^2)$$
$$(Q - W) - \tfrac{1}{2}(m_2^1 + m_2^2)(Q - V).$$

That is investment is worth while if the incremental benefits are greater than the capital and operating costs of the new capacity over the demand cycle *minus* any operating cost savings. With prices set equal to system short-run marginal cost in each sub-period the previous expression simplifies to:

$$\tfrac{1}{2}(m_1^1 + m_1^2)Q + \tfrac{1}{2}(m_2^1 + m_2^2)Q > K + 2b'Q.$$

Since capacity is assumed to be perfectly divisible the previous expression can be divided through by $Q$ to give:

$$m_1 + m_2 > k + 2b', \qquad \text{where } k = K/Q.$$

When capacity is optimal this inequality becomes an equality. Capacity is thus optimal when the sum of the short-run (system) marginal costs equals the long-run marginal cost of increasing output by 1 unit in both sub-periods. An alternative presentation of this result is that capacity is optimal when

$$(m_1 - b') + (m_2 - b') = k,$$

that is, when the system short-run marginal cost in each sub-period exceeds the short-run marginal cost of the new capacity (a quasi-rent) by an amount just sufficient to cover the 24-hour unit capital cost of the new capacity.

In Figure 4.4 the optimal capacity is assumed to be $0\bar{q}$, with optimal prices $p_1$ and $p_2$. The question to be considered is whether these prices would be optimal if storage was feasible. In order to consider this question let the cost of storing 1 unit of output for 12 hours be $c$. It then follows that storage will only be efficient if $p_2 - p_1 > C$, since storage will only be from the lower cost to the higher cost period. This inequality is satisfied in Figure 4.4.

The introduction of feasible storage will mean that production and consumption levels in each sub-period for the circumstances illustrated in Figure 4.4 will be different. The transfer of 1 unit of output from period 1 to period 2 will constitute a potential Pareto improvement (an efficiency gain) since $p_2 > p_1$. With feasible storage the optimum prices are $\overset{*}{p}_2$ and $\overset{*}{p}_1$, where $\overset{*}{p}_2 - \overset{*}{p}_1 = c$. Production in the off-peak period will be $0\overset{*}{q}_1$ and consumption $0\overset{*}{q}_2$, and in the peak period will be $0\bar{q}$ and $0\overset{*}{q}_3$ respectively, where $0\overset{*}{q}_3 - 0\bar{q} = 0\overset{*}{q}_1 - 0\overset{*}{q}_2$.

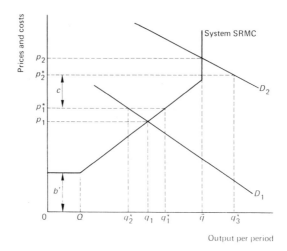

FIGURE 4.4

Notice that it is assumed that no wastage occurs while the product is in store.

The introduction of storage reduces the differential between both the peak and off-peak prices and their respective production levels but *increases* the difference between the consumption levels. It can be seen that if $D_1$ had been suitably drawn to the right the production peak could have been entirely eliminated, although compared with the non-storage case consumption differentials would have been increased.

An interesting result from the introduction of positive optimal storage is that optimal prices can be set solely by reference to cost conditions and thus without any information on demand functions,[19] as with the firm-peak case. It can be shown that for the assumptions made the optimal prices will be

$$\overset{*}{p}_2 = k + b' + \tfrac{1}{2}c$$
$$\overset{*}{p}_1 = k + b' - \tfrac{1}{2}c.$$

The peak price should exceed and the off-peak price fall short of long-run marginal cost by half the marginal storage cost.[20]

## 4.7 Evaluation

The application of the foregoing analysis to the determination of actual energy prices poses many problems. These include the definition of the appropriate number and duration of the pricing sub-periods, the measurement of the marginal costs, the measurement of the relevant demand functions, the allowing for transaction costs (and especially metering costs for electricity and gas) and the possible incorporation of equity considerations into prices. Various matters relating to these issues are taken up in this and the following two sections of this chapter.

Before we consider these issues it is important to note that the informational requirements of the firm-peak case are much less difficult to satisfy than those of the shifting-peak case. This is because only *cost* information is required if capacity is optimal.[21]

The essence of the previous analysis is that the resource costs of meeting an increment in demand will be different depending on (*a*) whether it can be met without requiring any additional investment, when it will involve only incremental operating costs, and (*b*) whether it will require additional investment, when it will involve both incremental capacity and operating costs.

If the number of sub-periods within the day is kept to two or three (and similarly for seasonal sub-periods) then for both electricity and gas in integrated supply systems the incremental costs within each sub-period will not be unique. Thus in an integrated electricity supply system which has a mix of plants of different vintages and technical types which are operated in merit order to enable demand per period to be met at least cost, the incremental system cost will vary with the precise timing of the demand increment. In these circumstances there may well be literally hundreds of marginal costs in any day, varying with the time, location and voltage level of the demand increment. Relevant questions thus concern the extent to which these cost differences should be reflected in the tariffs of particular consumer groups, how the price within each part of a tariff which is to reflect these cost differences should be calculated, and how the uniform price should be calculated for those consumer groups for whom it is decided for one reason or another that time-differentiated tariffs would be inappropriate?

## Simple versus Complex Tariffs

The determination of the extent to which tariffs should reflect cost differences, given the efficiency objective, depends largely on three factors. First, the costs of implementing and administering time-differentiated tariffs with different numbers of subdivisions, and for gas and electricity the most important item here will be the costs of meters with different numbers of dials and time switches. Second, the benefits derived from time-differentiated as compared with uniform tariffs. Third, the ease with which consumers can understand and thus react to the cost information contained in more complex tariffs.

The design of tariffs[22] must typically be considered at two levels, namely bulk supply and retail. The second and third of the points mentioned in the previous paragraph are unlikely to be important at the bulk supply level, and thus the design of these tariffs can concentrate on their incentive effects. At the retail level, since the costs of multi-dial meters do not vary with the amount of a consumer's consumption, it follows that the larger is the consumer the more likely it is that the incremental benefits of moving from single-price to time-differentiated tariffs will exceed the incremental costs. Thus there is likely to be a strong case for having such tariffs for industrial consumers and for the larger domestic and commercial consumers. Notice that consumers could be given the option of being on either simple or complex tariffs.[23] If they voluntarily choose to move to the latter this would constitute a Paretian improvement.

When determining the number of rates to include in a tariff account must be taken of its desired incentive effects. Predominant among these is the need to inform consumers when an increment in demand would involve the use of a relatively large quantity of resources and when it would involve relatively little. The greater the number the more accurately can the tariff reflect the structure of costs, but the greater will be the costs of implementing the tariff and the more difficult will it be for consumers to understand it. Whatever number is chosen, it will be necessary to group together times within a season or year when demand is approximately at the same level. The weighted average incremental cost must be calculated for each level.[24]

The problem of determining whether it is worth while changing consumers from simple to complex tariffs can be illustrated as follows. As before, let the objective be the maximisation of the sum of consumer and producer surpluses. Again assume that there are two 12-hour independent demand functions. Let there be a single-price tariff with the price set as a weighted average[25] of the marginal cost prices in the two sub-periods; call this price $s$. Let the constant marginal cost be $m$ in the peak period and $n$ in the off-peak period, as shown in Figure 4.5.

If the uniform price $s$ is charged over the demand cycle the peak and off-peak outputs would be $0q_2$ and $0q_1$ respectively, and the off-peak consumers would be cross-subsidising the peak-period consumers. If a time-differentiated tariff was used with price equal to marginal cost in each period, then the outputs would be $0q_3$ and $0q_4$ respectively. Thus if the tariff was changed there would be a net increase in output in

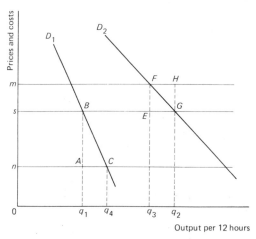

FIGURE 4.5

the off-peak period of $q_1 q_4$ and a net decrease in the peak period of $q_2 q_3$. In the peak period the industry's costs would be reduced by $0m \times q_2 q_3$, and in the off-peak period they would be increased by $0n \times q_1 q_4$. The consumers' willingness to pay for the change in their off-peak consumption would be $\frac{1}{2}(0n + 0s)q_1 q_4$ and for their peak consumption $\frac{1}{2}(0s + 0m)q_2 q_3$. Thus the net change would be:

$$[(0m \times q_2 q_3) - \tfrac{1}{2}(0s + 0m)q_2 q_3] + [\tfrac{1}{2}(0n + 0s)q_1 q_4$$
$$- (0n \times q_1 q_4)]$$

The net gain is thus equal to the sum of the areas $ABC$ and $FGH$ (=$EFG$). The change to the more complex tariff is thus worth while if the sum of these areas exceeds the additional metering and administrative costs per demand cycle associated with this tariff.

In the case of electricity, empirical evidence from England and Wales suggests that even for fairly large domestic consumers (with an annual consumption of 3000 kWh or more) the incremental benefits of moving to more complex tariffs will only exceed the incremental costs if the form of the tariff is kept relatively simple, such as containing a 'day' and 'night' differential.[26]

Time-differentiated tariffs for electricity, both bulk supply and retail, are in use in a number of countries, including France, Sweden, and the United Kingdom, and in part of Wisconsin in the United States. The National Energy Plan for the United States which was presented to Congress by President Carter proposed to make it mandatory for electric utilities to offer peak/off-peak tariffs to those consumers who were prepared to bear the associated metering costs.[27] In England and Wales the introduction of time-differentiated tariffs has had a substantial effect on the shape of the demand curve on the Central Electricity Generating Board's supply system.[28] It has been estimated that between 1960/1 and 1972/3 the use of such tariffs led to an improvement in the average daily load factor on that system from 72 per cent to 86 per cent, and to a reduction in the latter year of 4700 MW[29] in the peak demand on the average winter weekday.[30]

Similar effects have been reported by Electricité de France due to its Green Tariff. It has been estimated that the system peak demand has been reduced by approximately 800 MW.[31] In France about 20 per cent of domestic consumers are on time-of-day tariffs. In 1973/4 22 per cent of all sales to domestic consumers in England and Wales were made under the terms of time-differentiated tariffs.

## 4.8   Measures of Marginal Cost

The measurement of marginal costs poses many problems. But there is one supposed problem which it is important to dismiss. This relates to the variation of marginal cost with respect to the period over which it is measured. In general the measure of marginal cost will differ depending on whether it is measured with reference to a short or long period of time. There is a considerable literature in which economists debated the advantages of setting prices equal to either long- or short-run marginal cost. Apart from the fact that in equilibrium situations prices can simultaneously equal short- and long-run marginal costs many economists would now agree that this discussion misses the essential point. That is in the formulation of actual prices the period over which marginal cost should be measured should be determined by the length of the period to which the tariff will relate.[32]

For the energy industries there are a number of arguments in favour of this period being relatively long and generally measured in years. These include the fact that frequent changes in energy prices are expensive to administer and that it takes consumers time to adjust to them. In addition consumers of energy partly base their investment decisions for energy using complementary products on their views of the expected future prices of the different forms of energy. Thus prices relating to the long period are those required for the making of efficient investment decisions. But note that in times of inflation the *level* of charges may have to be changed fairly frequently; however, these changes should leave the *structure* of the tariff unchanged. The relevant question is thus, how do the energy utility's total costs vary with changes in its output during this chosen period?

Whatever time period is selected the measurement of marginal costs in the energy industries will often be complex because their operations need to be considered on a total systems basis (Section 4.6, p. 89). This is often clearest in the case of interconnected supply systems operated by electricity and gas utilities, but it also applies to coal.[33] In these cases the set of required marginal costs can be obtained from the system optimisation model which is used for investment planning.[34]

In the case of electricity with a least-cost objective this model will rank inherited plants in any day in order of merit. It can then be used to read off the system marginal operating cost once the precise timing of any demand increment for electricity is known. The investment problem is to choose that mix of plant with different cost characteristics, e.g. nuclear, oil-fired, coal-fired, hydro, which will enable forecast demand to be met at least-discounted cost. The solution of this problem must allow for the interdependence between the outputs of inherited and newly constructed capacity in any period, since the marginal operating costs of the latter plant will affect the number of hours for which the former plant is operated in each period.[35] If an increment in demand for electricity required the construction of new capacity then its associated marginal cost should in principle be calculated by taking the difference between the present value system costs associated with the new demand forecast and the equivalent figure for the previous demand forecast. This will give discounted marginal investment cost.[36] The marginal cost of an increment of peak load is thus equal to the sum of this discounted marginal investment cost, the system incremental running costs at that time and the incremental manning and maintenance costs. Notice that in the absence of spare capacity in all future periods the marginal costs of meeting a demand increment will vary with the length of the period of forewarning which is given to the affected utility. This is because there will be a minimum length of time required for the construction of the optimal (least-discounted cost) capacity. If the period of advance notice is less than this minimum period, then higher-cost non-optimal capacity must be constructed to meet the demand increment.

An interesting question which it is worth considering briefly at this point concerns the implication of the efficiency objective for the calculation of marginal generating costs in mixed hydro/thermal systems when the extra output would be produced by hydro capacity. In general it would not be correct to argue that since this capacity would have a zero fuel cost its marginal generation cost would be given by reference to its marginal manning and maintenance costs. The marginal generation cost should be calculated with reference to the marginal value of the water which would be used. In the absence of any alternative uses for this water (such as for irrigation) it should be valued in terms of the kilowatt hours which it could be used to generate. If an investment planning model is available then these figures can be obtained from it.[37] While it is not possible to generalise, the basic idea underlying the calculation of these figures can be explained by the use of two simple examples. First, if there is spare generating capacity and if the water was not used to generate electricity it would be spilled, then the marginal generating cost would be simply the hydro capacity's marginal manning and maintenance costs. Second, if there is spare capacity and if the water was not used to generate electricity it would be stored in a reservoir to replace potential thermal output, then the marginal generating cost is the marginal generating cost of this thermal plant plus the hydro station's marginal manning and maintenance cost (the resources which would be saved if the unit was not produced).

Considering the second of these examples it is clear that a prerequisite for the appropriate calculation of the marginal cost of the hydro station is that the marginal cost of the thermal station has been properly calculated. Since the justification for a set of prices based on marginal costs is the promotion of an efficient allocation of resources it follows that the relevant price base for the measurement of marginal costs is social opportunity cost. Because of the existence of external effects such as pollution, of transfer payments such as taxes and subsidies and of various governmental policies such as minimum wage laws and foreign exchange controls, market prices may be poor indicators of social opportunity costs. In that case the estimates of marginal costs should be based on the use of shadow or accounting prices. The use of these prices represents the formal introduction of second-best considerations into the analysis.[38]

Shadow prices can be calculated in various ways which differ with the objectives and constraints of a particular country. Our preference is to base them on border prices as suggested by Professors Little and Mirrlees.[39] For illustrative purposes the problem of calculating the marginal running cost of the thermal plant (which is assumed to be oil-fired) will be considered in a country such as the United Kingdom or Mexico where it is assumed that oil has recently been discovered. We assume that these discoveries will give an annual output (from the optimal depletion-rate decision) which is a small proportion of oil which is traded on the world market. The country is thus a price-taker. Let the marginal extraction costs of this oil be $2 a barrel and the investment costs be zero, so that it is worth exploiting at any price above $2 barrel. The basic question is how should the oil be valued?

The basic principles required to answer this question are the same as those used to determine the marginal cost of the hydro station. That is, the social opportunity cost of the oil must be calculated. Assume that the country places no restrictions on foreign trade and that its foreign-exchange rate is market determined. Under these circumstances the alternative to using the oil domestically is to export it at the world market price, which may be determined by a cartel as at present. If this price is $14 barrel for the grade of oil being considered, then although the marginal extraction costs are $2 this would represent the social opportunity cost of using the oil to generate electricity in the thermal station.[40] If there are restrictions on foreign trade this cost could be greater than $14 barrel. Thus suppose that there are quotas on a number of imports which could be released if some of the oil was exported. The value to consumers of a marginal relaxation of a quota restriction may well be considerably in excess of $14, and it would be their willingness to pay for the additional items which could be imported (measured in terms of border prices) which would represent the social cost of using the oil domestically rather than exporting it. It follows that the border price of oil may be taken as a minimum measure of its social opportunity cost to the power station.

The problem of the pricing of other domestic deposits of energy, such as coal or natural gas, can be determined using the same basic principles. In each case the social opportunity cost of the resource must be calculated along with a decision on the optimal depletion rate. This is illustrated for a particularly simple case involving the pricing of domestic gas and crude oil in Figure 4.6. Figure 4.6A shows the marginal extraction costs for natural gas deposits and $0q_1$ is the annual output limit set by the depletion-rate decision. Similarly Figure 4.6B shows the same data for domestic oil with the output limit of $0q_2$. It is assumed that natural gas and oil are perfect substitutes for each other (in fact natural gas is a better substitute for distillate fuel oil than for crude oil). In Figure 4.6C we have the demand curve for energy (measured over all potential uses of the energy, including its use as a feedstock in various industries), which is measured in therms, on the assumption that there are only the two fuels oil and gas. The figure also includes the border price of traded crude oil and the aggregate marginal extraction cost curve for the two fuels.

The optimal outputs are $0q_1$ and $0q_2$ for gas and oil respectively, while oil imports will be $q_m q_t$. In this case, with oil and gas being perfect substitutes, and the country being a

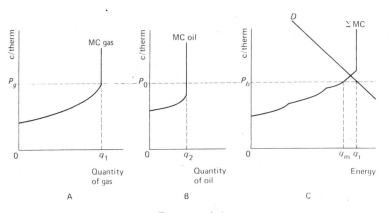

FIGURE 4.6

price-taker for oil, the function of the energy prices is to subdivide the total energy market between the three energy sources. The solution prices maximise the sum of consumers' and producers' surpluses.

In reality energy prices have to be determined allowing for various constraints set by the production technology, for objectives other than the efficiency objective, for the fact that different forms of energy are imperfect substitutes for each other; and that the relevant set of energy prices are delivered to the consumers' prices which must allow for transport and storage costs. Considering this last factor, the price $P_g$ in Figure 4.6A must be interpreted as *net* of pipeline, storage and other related costs, i.e. for offshore oil it is a beach price. As mentioned, different forms of energy are imperfect substitutes for each other (in Table 4.1, p. 78, the cross-elasticities would have a value of 1 if gas and oil, etc., were perfect substitutes for each other). In certain markets, such as domestic heating, natural gas has advantages over coal and oil in terms of its characteristics of ease of handling, cleanliness and convenience, and these markets are generally referred to as being 'premium markets'. In these cases the price of gas per therm will exceed that of oil or coal. $P_g$ can then be interpreted as a floor price for gas.[41]

## 4.9   Financial Targets

The analysis in this chapter has so far been concerned solely with the maximisation of social benefits minus social costs subject to the constraints given by the community's available resources and the state of technology. Many energy utilities, however, have to formulate their prices subject to an additional constraint, namely they are required to achieve pre-set financial targets. These may be expressed in a number of alternative ways; for example, as lump-sum cash targets or as rates of return. For regulated private utilities in the United States the targets are usually expressed in the form of *maximum* permitted rates of return.[42] For the nationalised energy industries in the United Kingdom the targets are expressed as *minimum* rate of return requirements. The incorporation of this additional constraint into the analysis means that the pricing problem is one of second-best optimisation.

For public sector energy utilities financial targets are equivalent to the imposition of a set of indirect taxes on their outputs. In such cases there are a number of reasons for preferring targets which are expressed in lump-sum rather than rate-of-return form. Predominant among these are that the resulting pricing rules are simpler and require less information.

The basic question which must now be considered is how should the pricing rules based on the efficiency objective be changed to achieve a financial target while minimising the resulting distortion to the allocation of resources? In considering this question we assume that the target has been expressed as a cash lump sum. A related question concerns the optimal way in which the targets should be set to the individual energy utilities.

The essence of the answer to the first of these questions can be gleaned from the following simple example. Assume that a financial target has been set to a gas utility which is selling gas on a time-differentiated tariff, peak and off-peak. For simplicity assume that the two demand curves are independent of each other and that the marginal cost of either product is unaffected by variations in the output of the other one. Finally, assume that over the expected range of prices the own price elasticity of demand for peak gas is equal to zero and for off-peak gas equals infinity. It is evident that in this case the whole of the target should be allocated to consumers of gas in the peak period. Since their demand is totally unresponsive to changes in price the implementation of this pricing policy will not change the allocation of resources (except at a second stage via associated income effects).

This example suggests that with the imposition of a lump-sum financial target the deviation of price from marginal cost should vary inversely with the own price elasticities of demand for the various products.[43] One interpretation of this is that the target should be allocated to the different products produced by a multi-product enterprise according to 'what the market will bear'. The efficiency loss associated with the imposition of the target will be minimised when the marginal efficiency loss per dollar contribution towards the target is the same for all products. Since the efficiency loss equals the difference between the price paid and the marginal cost of the product and the dollar contribution to the target is measured as the difference between the marginal revenue and the marginal cost of the product, this condition can be expressed as:[44]

$$\frac{P_1 - MC_1}{MR_1 - MC_1} = \frac{P_2 - MC_2}{MR_2 - MC_2} = \lambda,$$

where $\lambda$ is a constant with a negative value determined by the size of the financial target.[45]

In reality the various products of an energy utility are often substitutes for each other and thus the demand curves are interdependent. This means that the deviation of price from marginal cost must allow for both own and cross price elasticities. The optimising condition given in the last paragraph must still hold, although the measurement of the marginal dollar contribution must allow for both the direct and indirect effects of the price change on the utility's revenue. In the case of two goods we have:[46]

$$\frac{P_1 - MC_1}{MR_{11} + MR_{12} - MC_1} = \frac{P_2 - MC_2}{MR_{22} + MR_{21} - MC_2} = \lambda,$$

where $MR_{11} + MR_{12}$ measures the total effect on the enterprise's revenue of a change in output of good 1, and $MR_{22} + MR_{21}$ measures the total effect of a change in output of good 2.

A simple extension of the preceding analysis will enable us to answer the second question which was posed above. Given an aggregate financial target for a set of energy utilities the target for each individual utility should be set so that the required aggregate lump sum is raised with the resulting marginal efficiency loss being equal for each utility. If this was not the case then the individual targets could be changed so as to impose a smaller aggregate efficiency loss. For example, suppose that the ratio of marginal efficiency lost per dollar of surplus produced for a given allocation of the financial targets was 3 units in the electricity industry and 2 units in the gas industry. A reduction in the target for electricity by one dollar would lead to an efficiency gain of 3 units while an increase in the target for the gas industry by one dollar would lead to an efficiency loss of 2 units. Thus for the same aggregate surplus but a different subdivision there would be an efficiency gain of one unit. Only when the marginal efficiency losses per dollar of the lump-sum financial target were the same for all industries would it be impossible to make efficiency gains by marginal reallocations of the targets among the various industries.

The practical significance of this result is that the financial targets should be higher for those energy industries which face relatively inelastic demands over the current range of prices and relatively lower for industries with relatively elastic demand curves.

## 4.10 Equity

All pricing systems and changes in relative prices have distributional or equity implications. Thus, considering the change from a uniform to a complex tariff in Figure 4.5, p. 95, it can be seen that the peak-period consumers suffer a welfare loss while the off-peak consumers make a welfare gain. This raises the general question of what are the redistributional effects of peak-load tariffs? We know of no empirical evidence on this question. However, recognising the existence of these effects it is necessary to ask (i) whether they should be allowed for in the design of tariffs (a normative question) and (ii) if they should be then what redistributional schemes might be used to minimise efficiency losses.

At a first glance it would appear that at least in some countries or states the answer to (i) is 'yes'. For example, in the United States regulatory commissions are often required by statute to set rates which are 'just and reasonable' as well as non-discriminatory. However, this merely pushes the question back one stage to the definition of what is 'just and reasonable' and thus to the basic concept of equity. Now clearly there are many possible definitions of equity. To some equity is achieved if different consumers are charged a similar amount for a similar quantity of energy consumed, irrespective of any differences in the (marginal) cost of the supply of those services. This definition is obviously inconsistent with the efficiency objective. But equity can be defined to be consistent with that objective. Thus it can be defined so that all consumers placing the same costs on an energy utility should be faced with the same set of charges. (This definition of fairness was used in President Carter's National Energy Plan – see Chapter 9, Section 4.4.) The former type of definition involves an element of cross-subsidisation which is absent from the latter. A fundamental difference between these two approaches as to what constitutes equity is that the former one implicitly assumes that it is the responsibility of the energy utility to redistribute income while the latter definition assumes the opposite.

Utility prices, whether the utility is in the public sector or in the private sector but subject to regulatory control, are clearly a policy instrument which is available to a government to achieve equity goals. Some recent contributions to the literature have shown how the distributional effects of utility prices can, at least in principle, be measured.[47] While we fully recognise the potential distributional implications of different energy prices we have strong doubts as to the general desirability of using these prices to achieve equity objectives. For multi-product utilities operating in a second-best world the informational requirements are immense (this is because the marginal social utility of income must be estimated for all consumers). In addition, the use of this instrument involves transaction costs (which are usually omitted in the literature) which are in addition to those associated with the existing taxation and social security system. We can illustrate our views on this question by considering some aspects of the so-called 'inverted' or 'lifeline' tariffs.

The implicit assumption of the social welfare function which we have used in the derivation of the marginal cost pricing policy is that a dollar of benefit or cost has the same value irrespective of who receives or pays it. A set of energy prices which are related to marginal costs and weighted against those demands which are relatively price inelastic in order to realise financial targets may have what are considered to be adverse effects on the distribution of income. The increase in energy prices since 1973 has focused attention on this question in many countries because of their adverse effects on the budgets of low-income consumers.[48]

The response of some governments and observers to this situation has been to suggest (sometimes implicitly) that the energy industries should be treated as social services. Thus there have been suggestions that low-income consumers should receive a free allowance of, say, gas or electricity per period.[49] A more common suggestion has been that gas and electricity utilities should introduce what are known as inverted or lifeline tariffs.[50] These are tariffs in which there is no standing charge and in which the unit price is higher for large consumers than for small consumers. A simple form of such a tariff is to have an initial block of fuel (say 1500 kWh per quarter) priced below marginal cost and to charge all additional units at that rate which is required to enable a financial target to be met. Larger consumers would thus cross-subsidise smaller ones, and in effect they would by paying an indirect tax on each unit of their consumption which is charged at the higher rate.

The redistributional success of such tariffs requires that the level of electricity and gas consumption should be a good index of a consumer's income. Empirical evidence from a number of countries shows that this is not always the case and that the poor are often relatively large consumers of these products.[51] It is also worth noting that empirical evidence from the United States shows that average unit costs in electricity supply fall as consumption per consumer increases.[52]

This evidence is also relevant to the proposal that two-part tariffs should be replaced by uniform-rate tariffs. The latter are tariffs with no standing charge and under which every unit is sold at the same price. These tariffs also involve the cross-subsidisation of smaller by larger consumers.

A common feature of such schemes is that they involve the introduction of additional distortions into the price system, and in addition they are discriminatory in their treatment of large and small consumers. By opting for schemes which involve the cross-subsidisation of small consumers rather than granting them offsetting cash increases in social welfare payments[53] which they could spend how they liked, it could be argued that energy is being treated as a merit want and that consumers are not being treated as the best judges of their own welfare.[54] In market-type economies there is a general presumption that consumers do know what is best for themselves. It thus follows that a special case should be made out (and widely accepted) for treating energy as a good satisfying a merit want.

Note that the redistribution of income via the manipulation of relative energy prices involves the use of a set of hidden indirect taxes on the consumption of large consumers. The embodying price/tariff changes would probably not require the introduction of legislation and thus these implicit taxes would not be subject to the same scrutiny or assessed against the same criteria as other taxes. We prefer systems of income transfers and taxes which are explicit, since they can then be subject to economic and political debate.

Compared with policies which tackle poverty at its root causes by increasing old-age pensions and the like,[55] income redistribution policies based on the use of inverted tariffs and their like suffer from the disadvantage that they are non-selective. It is difficult to limit the increases in real income to the target group (small consumers who are rich will also benefit), and in addition these schemes give large consumers an incentive to try and beat the system if they are limited to some fuels. Thus if the price changes were limited to gas and electricity, with coal and oil prices being related to their marginal costs, then the change in relative prices would give large consumers an incentive to reduce their consumption of gas and electricity, which would bring them benefits as small consumers, and to increase their consumption of coal and oil which are now relatively cheaper. Since this adjustment may require consumers to change their stocks of appliances its effects would probably be long-term.

If the problem of income distribution is tackled by using energy prices as a target variable there is a danger that the underlying causes of poverty will be left untouched. Our preference is thus to deal with the question of income distribution (at least in developed countries) through the adoption of appropriate social security policies, and to relate energy prices, for given depletion-rate policies, to the efficiency objective. However, we also support the use of payment schemes for energy which spread the costs over the year as a series of relatively small payments (use of energy stamps, the making of constant weekly or monthly payments to the energy utility). This is because even if poverty could be eliminated through the adoption of appropriate social security policies family budgeting problems would probably mean that the payment of relatively large fuel bills (especially for gas and electricity in the winter when these are metered and billed quarterly) could still cause genuine hardship and distress.

## 4.11 Conclusions

In this chapter it has been argued that the set of relative prices in the energy sub-sector of the economy should be determined with reference to a common set of objectives for the individual energy utilities. Assuming an objective of an efficient allocation of resources it was shown that these prices should be related to their marginal costs. The term 'related to' rather than 'equal to' was used to allow for the costs involved in implementing and administering complex price systems, for the ease with which consumers can understand them, for the meeting of any financial targets, and for any significant price distortions which might exist elsewhere in the economy.

In conditions of limited information, risk and uncertainty (factors which have been under-emphasised in this chapter) the instruction to a multi-product energy utility to relate its prices to marginal costs will not determine a unique set of prices. The structure and level of these prices will depend to a large extent on the judgement of the price-setter. This will clearly make for difficulties for the implementation of any monitoring system which might accompany such a pricing instruction. Nevertheless, while there will always be room for argument it should be possible to determine whether prices are reasonably related to marginal costs. Simple tests of this may have to suffice, such as whether the measures of costs are forward-looking and based on forecasts rather than backward-looking and related to average historic costs.

# 5. ENERGY AND THE ENVIRONMENT

## 5.1 Introduction

In Chapter 4 it was argued that for an efficient allocation of resources energy prices should be related to their marginal social costs. In this chapter we consider one of the principal causes of the divergence of private from social costs in this sector of the economy. This is the external environmental costs associated with the production and use of energy. Some of these external costs are associated with the emissions of particular pollutants, such as sulphur dioxide and particulate matter, while others are associated with the loss of visual amenity, such as opencast (or strip) mining and overhead power lines. In addition some of these externalities are local in their effects while others involve transfrontier pollution.

This chapter is in seven sections. The various types of environmental damage associated with energy are considered in Section 5.2. Some evidence from various countries on the quantitative significance of various pollutants associated with energy is considered in Section 5.3. Some of the instruments which could be used for the control of these pollutants are discussed and analysed in Section 5.4. Various problems relating to the application of these instruments are considered in Section 5.5, and Section 5.6 considers some examples of control policy in practice. Finally, some conclusions are presented in Section 5.7.

## 5.2 Energy-associated Environmental Effects

For any particular type of energy environmental costs may arise at the stage of its production, transportation, or consumption, or at any two or all three of these stages. In this section the nature of the environmental effects associated with some forms of energy are considered briefly.

## Coal

The production and use of coal is a major source of environmental pollution in many countries. The generation of electricity using coal causes significant air pollution due to emissions of carbon dioxide, sulphur oxides, particulates and nitrogen oxides. It has been estimated that in the United States a 1000-MW coal-fired plant using 10,000 tons of coal a day would each year emit into the atmosphere about 8 million tons of carbon dioxide, 50,000 tons of sulphur oxide, 20,000 tons of nitrous oxide, and between 25,000 and 250,000 tons of particulate matter depending on how well stack emissions are cleaned before they are released.[1] Such a plant would also require the strip mining of approximately 1600 acres each year.

The control of sulphur oxide emissions from coal is difficult. It can be scrubbed from stack gases, but this tends to be expensive and reduces the thermal efficiency of the plant. The scrubbing of stack gases to remove sulphur dioxide produces sulphuric acid and unless this is collected it may be washed into water courses (the washing of the stack gases at the Fulham and Battersea power stations in London had to be abandoned because sulphuric acid was being washed into the river Thames). For the future the most promising method of reducing sulphur emissions is to change the combustion process from the use of pulverised-fuel firing to that of fluidised-bed. With fluidised-bed combustion ash is no longer emitted from the smoke stack and the volume of emitted gas is so low that cleaning it would be relatively cheap.

The use of tall stacks to disperse emissions from coal-fired power plants can lead to problems of transnational pollution. Thus, Britain has accepted responsibility for some of the airborne pollution falling as diluted sulphuric acid over Northern Europe. In Sweden acid levels in many lakes are so high that all vertebrate life has been killed. An O.E.C.D.

report blamed Britain for 40 per cent of the pollution which is slowly killing Scandinavian lakes, forests and vegetation.[2]

It is well known that air pollution is a contributory factor to a number of respiratory diseases. Lave and Seskin[3] compared total and infant mortality rates for 114 standard metropolitan statistical areas in the United States with a measure of the minimum concentration of either particulates or total sulphates in micrograms per cubic metre. Their results showed that a 10 per cent decrease in the minimum concentration of measured particulates would decrease the total death-rate by 0.5 per cent, while a 10 per cent decrease in the minimum concentration of sulphates would decrease the total death-rate by 0.4 per cent.

The adverse environmental effects associated with the mining of coal differ significantly between deep mining and strip (or opencast) mining. With deep mining the most important environmental problems are those of disease and death of miners. In the United States about 4000 miners a year get pneumoconiosis (black lung disease), and about 25 per cent of all coal miners become infected during their working lives. In Great Britain stringent control measures have reduced the incidence of this disease from 8.1 new cases per 1000 miners in 1955 to 1.9 per 1000 in 1967.[4] In the 1960s and early 1970s about 200 miners died each year in pit accidents in the United States. In the United Kingdom 59 miners were killed in 1975/6 and 38 in 1976/7.[5] Other environmental effects related to the deep mining of coal include those of subsidence, run-off of acid water and the visual disamenity and potential danger of coal tips.

The environmental costs associated with strip mining are more obvious than those of deep mining. Thus there are those of visual disamenity and of rainfall carrying acid into water courses. The restoration of land following strip mining can be expensive. Thus it has been estimated that in Maryland the simple restoration of this land cost $1075 per hectare in 1967 and had increased to $1248 per hectare in 1970. In England the full reclamation of such land (defined to include all associated engineering work, etc.) was estimated to vary between £4885 per hectare in the west Midlands to £9937 per hectare in the northern region in 1976.[6]

*Oil*

The main environmental effects associated with the extraction of oil relate to blowouts and seepage (especially for offshore drilling), and at the transportation stage to oil spills, tanker fires and explosions. In recent years there have been spectacular examples of each of these occurrences. Thus in January 1969 a leak occurred in a well owned by Union Oil at Santa Barbara, California. In the first hundred days of the leak over 78,000 barrels of oil spilled into the local ecosystem, at an estimated cost to society of $16.4 million.[7] In 1967 the oil tanker *Torrey Canyon* ran aground off the Scilly Isles, spilling over 30,000 tons of crude oil into the ocean, while in March 1978 the *Amoco Cadiz* went aground off Brittany, spilling over 200,000 tons of crude oil. The most serious accident to date involving a tanker explosion occurred in January 1979 when the *Betelgeuse* exploded and burnt out while unloading at the oil terminal at Bantry Bay in Ireland, killing 50 people.[8]

Most oil is not spilled in large accidents such as these, but in the succession of relatively small spills which appear to be an almost inevitable accompaniment to the production and transportation of oil. However, the ecological consequences of these small spills are generally much less severe than those of the large spills. For example, it is the large spills which can endanger whole species of fish by the destruction of hatcheries.

A recent interesting case concerning the environmental consequences of the route chosen for the transportation of crude oil concerns the Trans-Alaska Pipeline (TAP). This pipeline runs south across Alaska through two major mountain ranges and in its southern section it crosses the most seismically active region in North America. The pipeline terminates at the port of Valdez, from where the oil is shipped to final markets on the west coast. An alternative Trans-Canadian Pipeline (T.C.P.) was rejected. The relative merits of these alternative routes were the subject of a cost—benefit study undertaken by C. Cicchetti and A. Myrick Freeman.[9] These authors judged the T.C.P. to be environmentally less hazardous than the TAP. Using pre-1973 oil crisis relative prices they calculated that the net economic benefit (using a 10 per cent discount rate and a 25-year life) to the United States of the project had a median value of $1.48 per barrel for the TAP and of $1.65 per barrel for the T.C.P. (the net benefits would be even greater using post-1973 prices). They also calculated that the use of the T.C.P. route would give higher profits to the oil companies and higher royalties to Alaska. Thus on economic, environmental and profitability grounds the T.C.P. route appeared to be superior. Why then was the TAP route chosen? One very important reason appears to have been that the TAP route was the fastest to develop, and this was of crucial importance to B.P., which had an agreement to produce 600,000 barrels of oil a day by the end of 1977. While there were other reasons for the choice of TAP it is important to note that as the basis of the available evidence the chosen route was not consistent with the maximisation of U.S. and Canadian welfare.

## Nuclear Power

Nuclear power stations and their supporting plants (such as those for fuel reprocessing) are associated with a number of environmental effects. These stations, however, avoid many of the routine environmental effects associated with fossil-fuelled power stations. Thus there is no thermal pollution from the emission of carbon dioxide and associated long-term threat to the world's climate; there are no emissions of sulphur dioxide or fly-ash; no coal tip; and the power station is smaller.

Environmental dangers related to nuclear power vary from low-level radiation emissions to reactor explosions. The principal environmental effect is that of radiation. Unlike many of the environmental hazards associated with fossil fuels, radiation effects are mostly restricted to man. These effects are both somatic (to the individuals exposed to the radiation) and genetic (to their offspring). Because of these dual effects it is usual to measure radiation dose rates for both the bone marrow and for the reproductive organs. Table 5.1 shows the radiation doses received by an average member of the U.K. population in the mid-1970s.

This table shows that at the present time the radiation doses received as a result of nuclear power production in the

TABLE 5.1

*Dose Rates in the United Kingdom from Ionising Radiation*

|  | Bone marrow | Reproductive cells: genetically significant dose (G.S.D.) |
| --- | --- | --- |
|  | mrem/yr | mrem/yr |
| **Naturally-occurring** | | |
| From cosmic rays | 33 | 33 |
| From soil and airborne | 44 | 44 |
| Within the body | 24 | 28 |
| **Man-made** | | |
| Medical, diagnostic X-rays | 32 | 14 |
| Medical, radiotherapy | 12 | 5 |
| Medical, radioisotope use | 2 | 0.2 |
| Fallout from bomb tests | 6 | 4 |
| Occupational doses (not nuclear power) | 0.4 | 0.3 |
| Nuclear power industry | 0.25 | 0.2 |
| Miscellaneous | 0.3 | 0.3 |
|  | 154 | 129 |

N.B. The radiation dose is measured in terms of millirems (mrem).
     The allowable dose of radiation for a given organ of the body
     is usually measured in terms of rems, which is a unit of
     measurement which allows different sources of radiation to be
     compared

SOURCE: Royal Commission on Environmental Pollution, Sixth Report, *Nuclear Power and the Environment*, Cmnd 6618 (London: H.M.S.O., 1976) table 2, p. 16.

United Kingdom are small. But in early 1977 the U.K. nuclear industry was relatively small, with a total installed nuclear-generating capacity of 4600 MW. It has been forecast that in the year 2000 this installed capacity could be increased to 40,000 MW.[10] This large increase in installed capacity may mean that although at the present time radioactive emissions to the atmosphere are not a problem they could be in the future. The Royal Commission on Environmental Pollution, considering this issue, believed that a more systematic control of these emissions may be required in the future.[11]

One of the problems involved with the assessment of the dangers from radiation is that the doses which will cause particular medical effects are not known for certain.[12] The principal long-term effect of radiation is the induction of cancers. On this the Royal Commission concluded:

> A reasonable estimate of the number of fatal cancers that would be induced by a dose of 1 rem given to each of a million people would be of the order of 100, of which perhaps one quarter might be leukaemias. This means that radiation workers who receive an annual dose of 1 rem are running a risk of about 1 in 10,000 that they will eventually die of cancer as a result of each year's dose. This is approximately as dangerous as regularly smoking three cigarettes a week.[12]

Some of the main worries concerning the large-scale development of nuclear power relate to the problems of storing radioactive wastes; the security measures required to ensure that plutonium is not acquired by terrorists and criminals; the proliferation of nuclear weapons; and the probability of catastrophic nuclear accidents.

Nearly all the radioactive material produced by nuclear power stations is in concentrated form and must be isolated from man's environment. While the Royal Commission considered that present methods of treating wastes were adequate for the current size of nuclear programme it concluded that there should be no commitment to a large nuclear programme until a method for the indefinite containment of these wastes had been demonstrated beyond reasonable doubt. Since the half-lives of many of the fission products are very long (for plutonium 24,600 years) their production and storage raises important questions of inter-generational inheritance. In this context the nuclear process is best viewed as being irreversible and thus future generations will have no choice but to use resources monitoring and containing the nuclear waste of their predecessors.

The question of the probability of large-scale nuclear accidents is very controversial. Various estimates have been prepared. For example, Professor N. C. Rasmussen, who supervised the U.S. Atomic Energy Commission's *Reactor Safety Study*,[14] estimated that in the United States in 1969 the probability that a person would be killed by a motor vehicle was 1 in 4 x $10^3$, while the probability of being killed by a nuclear power plant was 1 in 5 x $10^9$ for persons living within twenty-five miles of a power plant.[15] Critics disputed this latter figure and pointed out that it applied to one person for one plant for one year, while by the year 2000 some hundreds of plants would be in operation. To date reactor accidents have been rare. Only one accident is known to have caused fatalities, three workers died at the U.S. Atomic Energy Commission's SL-1 plant in Idaho in 1961 (although there may have been a large-scale accident in the U.S.S.R. in the late 1950s). The only known large-scale escape of radioactivity to the surrounding ecosystem occurred as a result of a fire at the U.K. Atomic Energy Authority's No. 1 pile at Windscale in 1957.[16]

While it would be foolish to ignore the environmental dangers associated with the development of nuclear power it is important to remember that the use of fossil fuels is associated with their own set of dangers. Thus from society's point of point the crucial question relates to the acceptable trade-off between these fuels.

*Hydroelectric Power*

Reservoir-type hydroelectric power schemes are often regarded as being among the least harmful in terms of related environmental damage of all energy-producing projects. These schemes replace a single ecosystem with two new, and smaller, systems, one above and one below the dam. Hydro projects are often associated with favourable environmental effects; for example, with the development of good-quality fisheries (as with the development of trout fishing at the Hoover Dam in Nevada) and of recreation facilities. However, these projects are sometimes linked to substantial environmental costs. Thus, unless special 'fish ladders' are constructed they can prevent fish such as salmon from reaching their traditional spawning grounds. The conversion of a relatively rapidly flowing river into a much slower one into the lake behind the dam not only brings problems of siltation but may also increase the incidence of water-borne and related diseases. Thus malaria, bilharziasis, snail fever (schistosomiasis), liver fluke infections, etc., may increase in the region of the dam, especially in areas with tropical or sub-tropical climates. For example, schistosomiasis (which is a serious debilitating disease) increased dramatically in Egypt following the building of Aswan Dam.

Other environmental costs associated with the construction of storage-type hydroelectric projects may include the flooding of forests and the loss of wild-life sanctuaries; the deprivation of downstream farm land of its annual (fertile) silt deposits as a result of the reduction in the number and frequency of floods; and an increase in the loss of human and animal life as a result of accidents involving failure of the dam. With regard to the latter it has been estimated that in the United States about one dam gives way every year.[17] While most of these failures cause very few deaths[18] large failures come into the category of major disasters. Thus in 1962 two thousand people were killed in Longerone in Italy when a mountainside fell into the Vaiont reservoir and flooded the valley. Wilson and Jones have estimated that in the United States there is one major failure of the Vaiont size every fifty years.[19]

## Geothermal Energy

Geothermal energy is used in the generation of electricity in a number of countries, including Italy, Iceland, New Zealand, Japan, the United States, and the U.S.S.R. The electricity is generated using natural underground streams, as for example at the Geysers Power Plant in Sonoma County, California. Unlike fossil fuels, geothermal energy does not produce atmospheric particulate pollution. However, it is associated with relatively large emissions of hydrogen sulfide, sulphuric acid mist and arsenic. In addition it is associated with damage to vegetation and scenery, with the possibility of subsidence, and with the possibility of earthquakes due to the reinjection of the cold condensed steam into the fault zones.

The Japanese Environment Agency judged the adverse environmental effects associated with the development of geothermal energy to be sufficiently serious for it, in 1974, to reverse a decision made in November 1973 at the height of the 'energy crisis' to permit the sinking of fourteen geothermal wells after only six wells had been sunk.[20]

## Thermal Pollution

Energy production and consumption can be associated with the thermal pollution of both water courses and the atmosphere. The burning of fossil fuels has led to a significant increase in global concentrations of carbon dioxide. At the present time there appears to be no practicable way of controlling $CO_2$ emissions and thus these concentrations will continue to increase. They could have the effect of making the earth's climate much warmer through the so-called 'green-house effect'. This simply refers to the fact that an increase in $CO_2$ in the atmosphere, other concentrations unchanged, would increase the amount of infra-red radiation from the earth's surface which is reflected back. This would increase the earth's temperature (it has been estimated that every 10 per cent increase in $CO_2$ concentration will lead to a 0.2 per cent Centigrade increase in temperature) and could lead to significant climatic changes. At the present time there is no firm evidence that such an effect has occurred globally.[21] But various projections show that there is cause for concern about possible local climatic changes.

Heat release from the earth's surface is not uniform. In the early 1970s it was estimated that while the heat released over the United Kingdom was equivalent to 1 per cent of the solar input, in the London area it was 17.8 per cent. It has been estimated that if the fuel consumption in the London area was to increase to the level which would give the area a heat-release equivalent to a solar input of 50 per cent (which could happen in forty years' time), then this would certainly change the local climate substantially.[22]

## 5.3    Some Evidence on Energy-related Pollution Emissions

In this section we consider some evidence on the magnitude
and relative importance of various emissions of pollutants
associated with energy. For the sake of brevity attention will
be confined to emissions of air pollutants. These may be
measured in a number of ways. One measure which is frequently used is to calculate the amount of pollutants emitted
on a ton/day or ton/year basis. A problem with this measure
is that it does not take into account the assimilative capacity
of the environment. Thus it ignores the factors, such as wind
and climate, which govern the dispersion of pollutants once
they are emitted. In addition the weight of pollutants does
not measure their environmental impact. For example, in
terms of health hazard and visibility the environmental impact
of smaller particles is much greater than that of larger particles.

An alternative measure which is in common use is that of
ambient measurement. This attempts to measure the concentration of particular pollutants, such as sulphur dioxide,
particulates, in a given airshed. While this measure allows for
the assimilative capacity of the local environment it may be
defective in measuring the amount of pollution which is
absorbed by particular individuals. This is because it does not
allow for differences in population density and therefore for
the number of people exposed to a particular pollutant. The
effect on individuals is sometimes measured using human
'immissions', which is the quantity of pollution absorbed by
an individual in the course of his normal activity multiplied
by the number of people exposed to the pollution. This
measure is frequently used in the measurement of the effects
of radiation.

Air quality standards are often specified in terms of
ambient measures. Thus both the U.S. national and the
Californian air quality standards are defined in this way (see
Table 5.2). California introduced air quality standards in
1959, and it was the first American governmental jurisdiction
to set them. The Federal standards were set in 1971. They
act as goals for the air pollution control programmes. The
Federal standards are subdivided into primary and secondary,
the former being set to protect health and the latter welfare.

TABLE 5.2

*U.S. Ambient Air Quality Standards*

| Pollutant | Averaging time | Californian standards | Federal standards | |
|---|---|---|---|---|
| | | | Primary | Secondary |
| Photochemical oxides (corrected for $NO_2$) | 1 hour | 0.10 ppm | 0.08 ppm | 0.08 ppm |
| Carbon monoxide | 12 hours | 10 ppm | – | – |
| | 8 hours | – | 9 ppm | 9 ppm |
| | 1 hour | 40 ppm | 35 ppm | 35 ppm |
| Nitrogen dioxide | Annual average | – | 0.5 ppm | 0.5 ppm |
| | 1 hour | 0.25 ppm | – | – |
| Sulphur dioxide | Annual average | – | 0.3 | – |
| | 24 hours | 0.4 ppm | 0.14 | – |
| | 3 hours | – | – | 0.5 ppm |
| | 1 hour | 0.5 ppm | – | – |
| Suspended particulate matter | Annual geometic mean | 60 mg/m$^3$ | 75 mg/m$^3$ | 60 mg/m$^3$ |
| | 24 hours | 100 mg/m$^3$ | 260 mg/m$^3$ | 150 mg/m$^3$ |
| Hydrogen sulphide | 1 hour | 0.03 ppm | – | – |
| Hydrocarbons (non-methane) | 3 hours (6–9 a.m.) | – | 0.24 ppm | 0.24 ppm |

*Notes*: (*a*)  The Californian standard is not met when it is equalled or exceeded.
     (*b*)  The Federal standards, other than those based on annual averages or annual geometric means, are not to be exceeded more than once a year.
     (*c*)  ppm = parts per million.
     (*d*)  mg/m$^3$ = micrograms per cubic metre.

The Californian standards are solely concerned with health.

Total U.S. emissions in 1973 of major pollutants are given in Table 5.3.

Table 5.3 illustrates the dominant importance in 1973 of energy uses as the source of sulphur oxides, nitrogen oxides, hydrocarbons and carbon monoxide. The relative importance of energy uses as sources of air pollution is even more marked

## TABLE 5.3

### U.S. Emissions of Major Air Pollutants 1973

Millions metric tons per year

| Source | Carbon monoxide CO | Particulate matter PM | Sulphur oxides $SO_x$ | Hydrocarbons HC | Nitrogen oxides $NO_x$ |
|---|---|---|---|---|---|
| Motor vehicles (gasoline) | 73.00 | 0.78 | 0.22 | 11.76 | 6.84 |
| Motor vehicles (diesel) | 1.06 | 0.19 | 0.29 | 0.41 | 2.46 |
| Aircraft | 0.84 | 0.16 | 0.04 | 0.49 | 0.16 |
| Others | 1.09 | 0.03 | 0.10 | 1.55 | 0.19 |
| *Total transportation* | 75.99 | 1.16 | 0.65 | 14.21 | 9.65 |
| Coal | 0.63 | 5.62 | 20.47 | 0.15 | 5.35 |
| Fuel oil | 0.17 | 0.51 | 4.62 | 0.08 | 2.34 |
| Natural gas | 0.27 | 0.15 | 0.13 | 0.11 | 2.92 |
| Wood | 0.25 | 0.31 | 0.03 | 0.08 | 0.16 |
| *Total fuel combustion* | 1.39 | 6.65 | 25.47 | 0.47 | 10.96 |
| *Industrial processes* | 13.50 | 7.05 | 6.50 | 5.93 | 0.90 |
| *Solid waste disposal* | 3.28 | 0.57 | 0.08 | 0.96 | 0.16 |
| *Forest fires* | 1.84 | 0.22 | 0.0 | 0.32 | 0.05 |
| *Agricultural burning* | 0.99 | 0.28 | 0.0 | 0.33 | 0.03 |
| *Other* | 0.04 | 0.02 | 0.0 | 1.56 | 0.0 |
| *Total miscellaneous* | 2.87 | 0.52 | 0.0 | 2.21 | 0.08 |
| TOTAL | 97.0 | 15.9 | 32.70 | 23.78 | 21.75 |

N.B. Totals do not necessarily add up because of rounding errors.

SOURCE:: *1973 National Emissions Report* (U.S. Environmental Protection Agency, May 1976) EPA–450/2–76–007.

in some local areas of the United States. Thus in 1970 in New York City energy sources accounted for about 59 per cent of emissions of particulates, 100 per cent of sulphur oxides, almost 100 per cent of carbon monoxide and nitrogen oxides, and 82 per cent of hydrocarbons. Transportation accounted for 98 per cent of the carbon monoxide emissions and 66 per cent of the hydrocarbons. Stationary energy uses accounted for 47 per cent of the particulate and 95 per cent of the sulphur oxide emissions. The most significant non-energy source of air pollutants was solid waste disposal, which accounted for 40 per cent of particulate emissions.[23] Unlike most major cities of the world New York is character-ised by the almost complete absence of pollutants emitted by industrial sources.

The relative importance of transportation to air pollution, and in particular to the creation of smog, is well known in Los Angeles. The Los Angeles smog is quite different from that of London in the 1950s, and is known as 'photochemical' smog. It is formed by reactions in the atmosphere of hydro-carbons and oxides of nitrogen in the presence of sunlight. The majority of these reactants are contributed by the use of gasoline-fuelled motor vehicles.

In 1973 transportation was responsible for over 90 per cent of the air pollution problem in Los Angeles county. The principal emissions in 1973 by source are given in Table 5.4. The overwhelming importance of gasoline-fuelled vehicles in the emissions of hydrocarbons and carbon monoxide is clearly brought out in Table 5.4. In Los Angeles county the most important users of fossil fuels after motor vehicles are power stations. In 1973 they were responsible for 44 per cent of all the emissions of sulphur oxides, 13.8 per cent of partic-ulates and 10.2 per cent of oxides of nitrogen.[24]

Los Angeles county is in the South Coast Air Basin. Air pollution is not equally bad throughout the Basin, but exhibits marked locational differences. If the severity of pollution is measured by the number of days in the year that air quality standards are violated, then in 1970 the pollution was worse in the areas located along the San Gabriel Mountains (200 days or more) and was least serious in the areas near the coastline (less than 100 days).[25]

TABLE 5.4

*Air Pollution Emissions in Los Angeles county 1973*

Average Tons per day

| Source | Reactive hydrocarbons HC | Nitrogen oxides $NO_x$ | Particulates PM | Sulphur dioxide $SO_2$ | Carbon monoxide CO | Total |
|---|---|---|---|---|---|---|
| *Stationary sources* | | | | | | |
| Combustion of fuel* | — | 240 | 35 | 175 | — | 450 |
| Petroleum refining | 5 | 20 | 5 | 55 | — | 85 |
| Chemical processing | — | — | — | 60 | — | 60 |
| Organic solvent | 40 | — | 15 | — | — | 55 |
| Other | 30 | 20 | 10 | 30 | 10 | 100 |
| Sub-total | 75 | 280 | 65 | 320 | 10 | 750 |
| *Transportation* | | | | | | |
| Gasoline-powered vehicles | 690 | 775 | 40 | 30 | 7090 | 8625 |
| Diesel-powered vehicles | — | 20 | 10 | — | 20 | 50 |
| Aircraft | 5 | 15 | 15 | 5 | 160 | 200 |
| Ships and railroads | — | 25 | — | 10 | 20 | 55 |
| TOTAL | 770 | 1115 | 130 | 365 | 7300 | 9680 |

* Includes power plants

SOURCE: *1974 Profile of Air Pollution Control*, Air Pollution Control District, County of Los Angeles, p. 11.

This picture of the relative importance of energy use and production as a source of air pollution is confirmed by data from many other countries. Thus the famous London smogs, such as that of December 1952 which was estimated to have been the cause of more than 4000 deaths, were principally due to emissions of sulphur oxides and particulate matter from the combustion of coal and petroleum by industrial and household consumers. In Tokyo it has been estimated that in 1973 motor vehicles discharged 69.0 per cent of nitrogen oxides and power plants 16.3 per cent.[26] Unlike New York, industry (other than power plants) is a major source of air pollution in Tokyo, in 1973 it was responsible for 59 per cent of the emissions of sulphur oxides and 58 per cent of particulates.

### 5.4   Pollution Control Policy

There are a number of possible policy instruments available to governments wishing to reduce levels of pollution. These include the use of economic instruments, such as pollution charges or taxes, and regulatory instruments, like the laying down of emission standards, the control of firms' locations and the prohibition of the use of some fuels. Recognising the reciprocal nature of the external effects associated with energy production and use (that is, while the firm's economic activity unintentionally imposes uncompensated costs on third parties the removal of these costs imposes costs on the firm, as was stressed by Coase[27]), many governments appear in general to have accepted what the O.E.C.D. has termed 'the polluter pays principle' (P.P.P.).[28] That is, the polluter should bear the costs of measures set to achieve desired environmental standards, and these costs should be reflected in the prices of goods and services. There are, however, some important limitations to governments acceptance of this principle, and an important one refers to the measures which they initiate (or fail to initiate) to deal with trans-frontier pollution.

A problem which is common to the use of economic and regulatory control instruments is that air quality standards are generally specified using ambient measures, as in the United States, Japan and the German Democratic Republic. However, if charges and standards are related to the level of emissions then, if desired air quality standards are to be achieved, it is necessary to know the functional relationship between the level of emissions and their concentration in the air. We have previously noted that many factors, such as wind and climate, will affect this function and thus that it will be complex. The important point to note is that the determination of optimal standards, taxes, etc., will require the use of air diffusion models.[29] Henceforth in this section we shall assume that the form of this function is both known and specified.

### Standards and Charges for Pollution Control

Given sufficient information, desired levels of pollution abatement can be achieved, using either pollution charges or emission standards. To illustrate this we assume a perfectly competitive industry in which all firms are identical and which are operating in an airshed with a homogeneous air quality. Assume that one of the inputs used by the firms is coal and that this causes emissions of sulphur dioxide which cause unintended, uncompensated damage. For simplicity we assume that the pollution production function is proportional to the firm's level of output.

The top diagram in Figure 5.1 shows the perfectly competitive industry in long-run equilibrium producing output $0Q_1$ at price $0P_1$. The bottom diagram shows the associated discharge of sulphur dioxide and the aggregate marginal control cost (MCC) and marginal damage cost (MDC) functions (which on the assumptions made will be the same for each firm). The MCC curve is drawn on the assumption that the marginal cost of reducing emissions increases monotonically with the level of planned reduction, while the MDC curve is drawn on the assumption that the marginal damage associated with the level of emissions increases monotonically with the level of emissions (some problems relating to the measurement of damage cost functions are considered in Section 5.5).

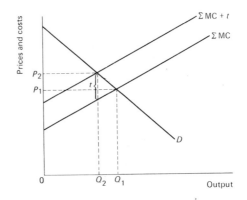

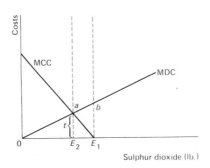

FIGURE 5.1

The benefits to society of pollution control consist of averted costs, and thus benefits are potentially measured by the MDC curve. It is clear that the level of output $0Q_1$, with sulphur dioxide emissions of $0E_1$, is not socially optimal. A marginal reduction in output from $0Q_1$ produces positive net benefits to society since MDC > MCC. This is true for all reductions in output to $0Q_2$ and of emissions to $0E_2$. This latter is, for the assumed functions, the optimal level of pollution. The industry can be induced to reduce its output to $0Q_2$ in a number of ways. One is to impose a Pigovian tax at the rate $t$ on each pound of sulphur dioxide emitted.[30] This will give the industry a new aggregate marginal cost curve MC + $t$. The profit-maximising output is then $0Q_2$, which gives the socially optimal level of pollution $OE_2$. At the optimal level of pollution the marginal cost of reducing pollution concentration is equal for all emitters. (In the case considered this result follows directly from the assumption that all firms are identical, but it can easily be shown that it is a necessary condition for the achievement of the desired level of air quality at least cost.)

To set the optimal tax information is required on both the MCC and MDC functions. If this information is available it can be used to set the optimal emission standard. If there are $n$ firms then the standard for the individual firm will be $OE_2/n$.

Although the same reduction in the level of emissions can be achieved, using either of these policy instruments, there are some important differences in their associated effects. First, their distributional effects are different. The use of the tax involves each firm in making a payment of $t$ multiplied by the number of pounds of $SO_2$ which it continues to emit. In contrast the use of the standard involves no such payment. Second, the implementation and enforcement costs of the two policies will be different. While both require the monitoring of emissions and the use of legal powers (in the one case to enforce the standard and in the other to ensure that due tax payments are made) the tax solution also requires the use of resources for the collection of the appropriate taxes.[31] Note that in terms of Figure 5.1 the case on efficiency grounds for the introduction of either the tax or the standard requires that in each case their associated non-marginal implementation and administrative costs be less than $abE_1$ (the net increase in social benefits as a result of the abatement of pollution).

*Policy in the Absence of Data on MDC*

In general for emissions associated with energy there is a lack of data on marginal damage cost functions.[32] In that case, as Baumol and Oates[33] have argued, 'politically acceptable' rather than optimal levels of pollution are likely to be the objectives of public policy. They have showed that the least-

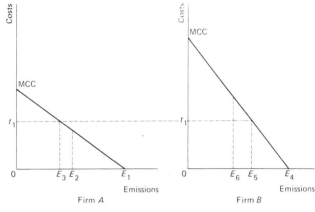

FIGURE 5.2

cost method of attaining a particular standard will be a tax per unit of pollution equal to the long-run marginal costs of abatement at the 'acceptable' level.

Figure 5.2 shows the MCC curves for two firms operating in a given airshed. The two policies to be considered are the imposition of a uniform tax or a uniform standard, both of which are set so as to reduce total emissions by 50 per cent. The imposition of the standard requires firm $A$ to abate emissions by $E_1 E_2$ and firm $B$ by $E_4 E_6$. With this solution the marginal cost of reducing emissions is different for the two firms, and is lower for firm $A$ than for $B$. This means that the total cost of reducing emissions by the required amount can be lowered by increasing the abatement in $A$ from $E_2$ to $E_3$, and reducing that in $B$ from $E_6$ to $E_5$. At these emission levels the total cost of achieving the desired abatement is minimised and this is the solution achieved by imposing the uniform tax rate $0t_1$. This constitutes the usual economic case for the claimed superiority of the pollution charge over the standard.[34]

## Technological Change

Pollution charges are also superior to regulatory standards in terms of the incentive which they give to firms to seek new methods for reducing emission control costs. In Figure 5.3

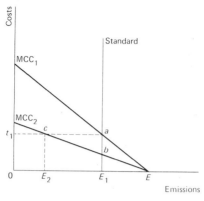

FIGURE 5.3

we show an initial standard at the level $E_1$, and an equivalent emission-reducing tax rate of $t_1$. The original and new marginal control cost functions are designated $MCC_1$ and $MCC_2$ respectively. With the standard the firm's incentive to seek the cost-reducing method of control is limited to the reduction in control costs shown as *Eba*. With the tax, however, this incentive is given by the sum of *Eba* and $E_2 caE_1$ (the latter being the saving in tax payments). Not only is the incentive to reduce emission control costs stronger with the tax, but in addition its use leads to a reduction of emissions — to $E_2$ — a result which does not occur if the standard is used.

## Locational Factors

A major deficiency of the preceding analysis is that it assumes that the quality of the airshed is the same for all pollutant-discharging firms. In reality, as we have seen in the case of the Los Angeles airshed, this is not the case. There will be important locational differences in the damage costs associated with particular emissions. Thus they will be different in urban and rural areas, and for firms in urban areas they will be different for those areas which are subject to frequent temperature inversions (such as Los Angeles) from those which are not. We will now extend the previous analysis to allow for local variations in damage costs.[35]

In Figure 5.4 are shown two firms with identical MCC

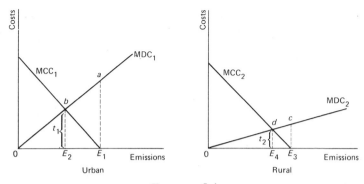

FIGURE 5.4

functions but different MDC functions due to locational factors. One firm is assumed to operate in an urban area and the other in a rural area. The optimal level of pollution reduction in the urban area is $E_1 E_2$ and in the rural area $E_3 E_4$. These reductions can be achieved either by the application of the unit taxes $t_1$ and $t_2$ on the urban and rural firms respectively, or by imposing the individual standards at the levels $E_2$ and $E_4$.

We have previously noted that standards and taxes have different distributional implications. Considering Figure 5.4 it is important to note that a system of differential taxes also has some (unintended) distributional consequences. To see this assume that all firms are perfectly competitive and that before the imposition of the corrective taxes the prices of all inputs, and their quality, was identical for all firms. The introduction of the system of differential taxes is equivalent to changing the relative price of land in the two areas in favour of the rural area. This will result in the latter land earning an economic rent, since firms will take the tax into account when making their location decisions. Thus as a result of the government tax policy to internalise the external costs there will be an unintended redistribution of income towards the owners of rural land.

## Timing of Control Measures

The preceding analysis has assumed that governments are only interested in the quality of the environment and not in the date at which a particular quality target will be achieved. But this is clearly incorrect; in general governments (as with the United States and Japanese Governments, in setting the emission standards for cars) are interested in both of these variables. The introduction of the timing of abatement means that there will be two policy targets, quality and timing. Their achievement will require the use of two policy instruments, since with a single-policy instrument the Government can control either the amount of abatement or the time of abatement, but not both.[36]

## 5.5 Problems of Application

Whether it is decided to control pollution by the setting of regulatory standards or the use of pollution charges there are a number of problems involved in formulating and implementing the chosen policy. These include problems associated with the measurement of emissions, the measurement of damage-cost functions and the setting of optimal taxes or regulations when energy production or use is associated with the emission of a number of pollutants.

### Measurement of Emissions

There are a number of problems associated with the measurement of emissions. The use of either standards or charges requires that changes in environmental quality can be related to specific discharges; that means are available for the accurate measurement of emissions; that the chosen control method can be easily understood by the affected parties; and that it is reasonably cheap to implement and administer. In practice problems occur with each of these requirements.

In many cases there is incomplete scientific knowledge of the function relating changes in environmental quality to particular emissions. In some areas little information is available on the environmental impacts or effects of particular emissions. Examples include the long-term climatic effects of carbon dioxide emissions and the somatic and genetic effects of radiation. The absence of complete information on the physical effects of particular emissions clearly precludes their accurate expression in monetary terms.

At the present time not only is information often incomplete on the reaction function between emissions and changes in environmental quality, but, in addition, available equipment for measuring emissions is sometimes both expensive and inaccurate. Thus it has been estimated that equipment for the automatic measuring of $SO_2$ emissions has typically an error of 15–20 per cent.[37]

Even if the necessary equipment was relatively cheap and accurate there could still be a major problem of implementation in the case of pollution caused by a large number of small emitters, as with the infamous London and Los Angeles smogs. At the time of the 1952 London smog about 43 per cent of the smoke and 18 per cent of the sulphur dioxide in the air in the United Kingdom came from chimneys of houses.[38] The control of these emissions, using the tax instrument, would require the installation of an emission meter in each house, its periodic reading and billing of the occupant. Clearly in this case the transactions costs associated with the use of the tax, and the problems which individual emitters would have in understanding it, mean that the tax solution is almost certainly non-optimal.[39] Similar problems can arise with the use of 'reasonable' standards, although they can be set in such a way that they are both easier to understand and involve lower transactions costs than the use of the tax instrument. This is the case, for example, when the standard for a particular emission is set at zero. This was the solution adopted in the United Kingdom for the control of smoke emissions. The 1956 and 1968 Clean Air Acts gave local authorities powers to create smoke-control areas (smokeless zones) in which it is an offence to emit smoke from any chimney unless the premises concerned have been exempted.[40]

The fact that there are a large number of small polluters does not necessarily mean that the tax solution is inapplicable. Thus an ingenious scheme for taxing smog-contributing motorists has been put forward by economists at the Rand Corporation.[41] One version of their proposed smog tax involved the periodic testing of cars and the assignment of a smog rating which could be indicated by a seal attached to the car. This would be used to determine the tax which the motorist would pay when he purchased gasoline. The tax could be varied by area or airshed.

This proposal raised an important question relating to the policing of the tax revenue collected by garages. Presumably to avoid fraud they would have to issue registered smog tax tickets (which could be monitored) to motorists.

## Damage-cost Functions

The determination of optimal pollution taxes or standards requires data on both marginal damage-cost and control-cost functions. The estimation of damage-cost functions involves two steps. First, the estimation of the physical effects of the pollutants, and second their valuation in monetary terms. Many problems are involved at each of these steps. Many of the problems at the second stage arise from the absence of markets for many environmental goods; for example, property rights do not exist for items such as clean air. In addition the relevant measure of damage costs is the present value of the damage over all future time periods. This means that many of the problems which were discussed in Chapter 3 on depletion rates are equally applicable to the problem being discussed in this section, namely the absence of a complete set of futures markets, risk and uncertainty, and the choice of the optimum discount rate. Some of the difficulties in measuring damage-cost functions will now be considered briefly.

The physical damage resulting from a change in emissions and ambient air quality will depend on the assimilative capacity of the environment. When it occurs this damage may take many forms.[42] Thus there may be adverse effects on climate and weather due to an increase in the concentration of carbon dioxide; adverse effects on natural resources, such as the loss of recreation benefits due to visual disamenity caused by strip mining, or the loss of agricultural output due to an increase in soil acidity; adverse effects on flora and fauna, and on materials such as the corrosion of buildings; and, finally, adverse effects on human health.[43] Physical damage may be separated by both time and space from the source of an emission. Examples include the possible genetic effects resulting from exposure to radiation and sulphur dioxide emissions from industrial firms in the Ruhr (West Germany) which, given certain weather conditions, spread to Scandinavia where they cause air pollution and increased acidity in lakes. The second of these examples raises a familiar problem in cost–benefit analysis, namely where to draw the boundary for the costs and benefits which are to be counted. Typically this is drawn on national lines. Thus the solution to problems of trans-frontier pollution requires international co-operation. Other problems relate to the choice of the units in which physical damage is to be measured and allowing for risk and uncertainty.

Assuming that the relevant objective is an efficient allocation of resources, that the initial distribution of income and wealth is judged to be acceptable, then possible policy changes can be evaluated in terms of potential Pareto improvements. In that case damage-cost functions should be valued in terms of Hick's compensating variations.[44] The compensating variation measures the money transfers to and from an individual which will keep him in his initial welfare position. It can be defined as either the minimum sum of money required to compensate an individual for a change which affects him adversely, or as the maximum amount which he would be willing to pay to derive the benefits from a change which affects him favourably.

Unless income effects of price changes are equal to zero valuation in terms of compensating variations requires the use of compensated demand curves (demand curves which are measured solely in terms of substitution effects).[45] Due to data limitations in empirical work on the estimation of damage-cost functions it may be impossible to estimate compensated demand curves whence it would be necessary to work with ordinary demand curves. But this need not lead to a great loss of accuracy; it all depends on whether or not the income effects associated with environmental changes are significant. If a large number of individuals are each affected in a small way by the damage caused by a pollutant then it is unlikely that income effects will be significant. On the other hand the smaller the number of persons who are affected significantly by a change in the environment the less severe are the estimation problems associated with compensated demand curves.

A change in environmental quality may alter the probability of certain events. Thus an increase in the concentration of sulphates in the atmosphere may increase the probability of a person dying from various respiratory diseases. The damage-cost estimate should then be in terms of the sum required to compensate the individual to make him indifferent between the two probability distributions. To state this principle is to raise many problems. A well-known one relates to the compensation required to offset an increased probability of death, and this relates to the problem of valuing life. In the present context this valuation problem is complicated by the fact that many individuals will not perceive the relationship between an increase in the emissions of particular pollutants and an increase in the probability of their death in each future year of their expected life. Clearly what is not perceived cannot be valued in terms of compensating payments. A pertinent example of this relates to the risk of catastrophic accidents from the construction of nuclear power stations, for which the probability distribution is not known.

Estimates of air pollution damage costs from all sources (not merely energy) are available for the United States and are given in Table 5.5. This table must be interpreted with great care. Many of the damage-cost categories are not additive. Thus changes in property values almost certainly include damage to the fabric of domestic properties, which is included under materials, and may also include part of the soiling cost. With available data it is not possible to determine to what extent some costs are included in other costs. Further, changes in property values have to be interpreted with very great care if they are to be used as measures of the monetary value of damage. The principal problem is to determine whether observed changes in market prices measure either the compensation property owners require to accept air pollution or their willingness to pay for the removal of air pollution.[46]

*Problems of Multiple Emissions*

The analysis of Section 5.4 was in terms of a particular

TABLE 5.5

*Damage Costs due to Air Pollution in the United States 1970*

$ millions (range)

| Type of damage | Particulates | $SO_2$ | $NO_2$ | Oxidants | CO | Total |
|---|---|---|---|---|---|---|
| Health | 8–41 | 1–4 | 0–1 | 44–221 | 9–44 | 62–311 |
| Materials | 169–767 | 215–972 | 52–236 | 254–791 | 30–134 | 720–2900 |
| Plants | 10 | 4 | 2 | 116 | 0 | 132 |
| Animals | 1–3 | 0 | 0 | 4–11 | 0–2 | 5–16 |
| Property values | 300–1765 | 126–745 | 47–275 | 67–399 | 19–114 | 559–3298 |
| Soiling | 519–2077 | 0 | 0 | 0 | 0 | 519–2077 |
| Total | 1007–4663 | 346–1725 | 101–514 | 483–1538 | 58–294 | 1995–8734 |

SOURCE: T. E. Waddell, *The Economic Damages of Air Pollution* (Washington, D.C.: U.S. Environmental Protection Agency, 1974).

activity generating emissions of a single pollutant. But, as was pointed out in Section 5.2, energy production and use is generally associated with the simultaneous production of a number of pollutants. This factor must be incorporated into the preceding analysis.

If equipment can be assumed to be available to measure each of the emissions then, providing the relevant marginal damage and control functions are known, the optimal tax can be set on each emission.[47] This can be illustrated using Figure 5.4 (p. 132), which must be interpreted as showing two separate emissions from a particular activity. If the tax policy is to be optimal the sum of the associated transaction costs (which include the costs of the required metering equipment) must be less than the sum of the areas $E_1 ab$ and $E_3 cd$ in Figure 5.4.

Suppose, however, that the sum of the transactions costs is greater than the sum of these areas. This would suggest that the optimal policy is not to control some of the emissions, since although the marginal benefits from control exceed the marginal costs, the total benefits from control are less than the associated total costs. However, the question is posed as to whether there might be a single weighted average tax for the various emissions for which the associated transactions costs are less than the sum of the net benefits arising from the imposition of the tax. If the various emissions were always (for all production functions) produced in fixed proportions then the optimal policy would be to monitor the emission which could be measured most cheaply and to levy the tax on it, assuming that the transactions costs are less than the sum of the associated net benefits. Technological change which would alter the emission proportions would require the recalculation of the optimal tax. The principal problem with the suggested weighted average tax is that information is required on all the marginal damage and control-cost functions for its calculation.

## 5.6  Policy in Practice

We have been unable to find examples of pollution taxes which are applied to emissions into the atmosphere (they are, however, used in some countries for emissions into water courses).[48] Two examples which appear at first sight to involve pollution taxes on closer examination prove not to be examples of the type of taxes considered earlier in Section 5.4. These examples relate to the control of sulphur emissions in Norway and the Netherlands.

Since 1971 Norway has had a tax on the sulphur content of fuels derived from petroleum. This tax has two objectives, first to reduce $SO_2$ emissions and second to raise revenue. The tax does not take account of any treatment which firms may undertake to reduce sulphur emissions from fuel used, does not vary regionally with pollution concentrations in the atmosphere, and is not levied on emitters but on oil suppliers. The tax consists of two parts: first, a basic charge of 1 öre per litre of oil regardless of sulphur content (revenue-raising), and second a charge based on sulphur content which varies from zero for a sulphur content of up to 0.5 per cent by weight to a charge of 1.2 öre per litre for a content of 3.0 to 3.5 per cent by weight.[49] The first part of the tax is not applicable to all users of oil; for example, power stations are exempt. It is clear that this tax, since it is not levied on emitters and does not take account of treatment undertaken to reduce emissions, is at best a third-best approximation to the type of tax discussed earlier in this section. However, it should be borne in mind that Norwegian policy for abating $SO_2$ emissions relies mainly on direct regulation. This takes the form of setting product standards for the sulphur content of different fuels according to the purpose for which they are used.

The control of $SO_2$ emissions in the Netherlands is also largely based on the setting of product standards and the use of licensing procedures.[50] The former of these are aimed principally at reducing total $SO_2$ concentrations, while the latter are aimed at reducing local concentrations. The Netherland's Air Pollution Act stipulates that the costs of implementing this legislation should be financed by a charge, which came into effect in 1972. This charge is related to the type of fuel and not to its sulphur content. It plays no incentive role.

Many countries attempt to control sulphur oxide emissions by setting permitted standards for the sulphur content of

TABLE 5.6

*Permitted Sulphur Content of Fuels in France*

| Fuel | Prior to 1974 % | 1974 % | 1978 or 1980 % |
|------|------|------|------|
| Domestic fuel oil | 0.70 | 0.55 | 0.3 |
| Light fuel oil | 2.0 | 2.0 | cancelled 1975 |
| Heavy fuel oil, No. 1 | 2.0 | 2.0 | 2.0 |
| Heavy fuel oil, No. 2 | 4.0 | 4.0 | 4.0 |

fuels. The maximum sulphur content by weight for different liquid fuels in France are given in Table 5.6.

Policy in Japan emphasises the control of sulphur content of heavy fuel oils. Similarly, control of sulphur emissions in the United States has relied on the use of regulations and not on the use of pollution taxes. Thus we have previously noted that in Los Angeles county power stations both have been and are an important source of emissions. They have been controlled via the enactment of certain regulations. Before 1967 these concentrated on the substitution of natural gas for high sulphur fuel oil. In 1968 a rule was enacted (Number 62.2) requiring all users of fuel oil to burn low-sulphur fuel oil whenever natural gas was not available. At that time oil from Indonesia and Alaska was becoming available in substantial quantities, and its sulphur content was 0.5 per cent by weight, or less.

Although the possibility of a smog tax on cars has been considered (p. 134), those countries which attempt to control emissions from mobile sources rely on a range of non-price control measures, such as the setting of vehicle emission standards (as in the United States and Japan) and the use of controls aimed at reducing vehicle miles travelled (V.M.T.). A reduction in V.M.T. would not only cause an improvement in air quality (see Table 5.7), but in addition it would reduce congestion and aid energy conservation. The range and type of measures which have been proposed to reduce V.M.T. can be illustrated with the Short-Range Transportation Plan which was adopted on 11 April 1974 by the Southern California

TABLE 5.7

*Average United States Emissions per Mile Travelled*

(in grams per mile)

| Pollutant | 1965 | 1970 | 1971 | 1972 | 1973 |
|---|---|---|---|---|---|
| Carbon monoxide | 89 | 78 | 74 | 68 | 52 |
| Hydrocarbons (exhaust) | 9.2 | 7.8 | 7.2 | 6.6 | 6.1 |
| Hydrocarbons (crankcase and evaporation) | 5.8 | 3.9 | 3.5 | 2.9 | 2.4 |
| Nitrogen oxides | 4.8 | 5.3 | 5.4 | 5.4 | 5.4 |

SOURCE: Environmental Protection Agency, *Monitoring and Air Quality Trends Report* (1972, 1973) table 4–13.

Association of Governments (S.C.A.G.).[51] This plan emphasised the development of incentives for the formation of carpools and for vehicle users to transfer to buses. Thus it involved an expansion of the bus system and the granting of preferential treatment for buses and carpools on freeway lanes and major arterial roads with the aim of reducing journey times.

## 5.6   Conclusions

In this chapter we considered the nature and quantitative significance of various pollutants associated with the production, distribution and use of energy. It was seen that such pollutants are pervasive and sometimes involve problems of transfrontier pollution. The relative merits of regulations and pollution taxes as control instruments were analysed[52] and the types of control policies which are used in practice were considered briefly. It was noted that at the time of writing no country appeared to be using taxes for the control of emissions or residuals associated with energy.

There are a number of problems which must be overcome before optimal taxes or regulations can be used as control instruments. Many of these relate to the problems of measuring and valuing the relevant damage-cost functions. These include problems of relating emissions to particular dischargers, problems of determining the functional relationship between the ambient air quality in an airshed and particular emissions, the problem of determining optimal taxes or regulations when the pollution production function is multiproduct, the problem of valuing emissions when the effects are irreversible and will mainly effect future generations, and so on.

To recognise these problems is not to argue that energy prices should be related to their marginal private costs and that the social costs associated with environmental damage should be ignored. Rather it is to argue for more research into the estimation of the relevant damage-cost functions and for the implementation of second-best pricing policies based upon such information about the relevant functions as is available. This is a policy area in which there appears to be no practical alternative to the adoption of piecemeal optimisation policies.

# 6. FISCAL INSTRUMENTS OF ENERGY POLICY

## 6.1 Introduction

An elementary theorem in the theory of public policy indicates that if the Government wishes to achieve certain values of given 'target' variables, for example a particular level of employment or size of balance-of-payments surplus, it must have at its disposal at least as many different policy instruments.[1] In the field of energy economics this restriction is particularly important because of the extreme complexity and range of objectives which are frequently involved. In Chapter 3 we noted some of the problems in efficiency such as information externalities, and ill-defined property rights (i.e. common pools) which could lead to sub-optimal levels of exploration and too rapid a rate of depletion of known resources. Chapter 4 made reference to the effects on income distribution of energy pricing policies, while in Chapter 5 the problems of pollution externalities which arise from energy use were examined. The development of energy resources may have important consequences for employment and regional policy, as well as for other macro-economic aggregates such as the balance of payments. Governments are also observed to be interested in the extent of foreign control over energy resources and the implications for national security of relying on various sources of supply. Exploitation of energy resources may also enable companies to receive large elements of economic rent which governments often judge should more properly accrue to the state.

This variety and breadth of policy interest inevitably makes energy policy a very complicated problem in public finance. As an example, consider a tax introduced to control the rate of depletion of an indigenous energy source. In certain circumstances it may succeed only at the cost of encouraging imports which now replace domestic supplies. This result would be counter to any balance-of-payments or national security objective. A tariff or quota might then reduce imports, but the resulting protection and higher domestic prices could run foul of income distribution objectives. They would also, *ceteris paribus*, increase the profitability of domestic production and thwart any attempt at taxing excess profits. Further there is no assurance that the prevailing price would be appropriate on environmental grounds. The resource might have a high 'user cost' but rather low associated environmental costs. This is the type of dilemma which occurs in the case of natural gas, which is a relatively 'clean' fuel.

In the following sections it is assumed that pollution problems are tackled by fiscal or other means as described in Chapter 5, that production externalities stemming from resources held in common are resolved by field unitisation or, where appropriate, international agreement,[2] and that aggregate monetary and fiscal policies are capable of solving problems of macro-economic adjustment. Two major issues are investigated: the effects of various tax instruments on the depletion rate, and the various methods available for taxing economic rent.

The latter problem has taken on increasing significance in recent years with the higher prevailing prices of energy resources creating windfall gains for many producers. Whether such gains 'ought' to be taxed at higher rates than apply to normal income is entirely a matter of equity. When the resources concerned are located offshore under the continental shelf or are discovered on federal land the usual judgement is that any economic rent should accrue to the state. Where resources are discovered on private land opinions vary. In the United Kingdom many mineral rights are vested in the Crown whereas in the United States, where property rights are perhaps less restricted, gains to private owners are treated as capital gains for tax purposes. Apart from considerations of what is 'just', the taxation of economic rent has the theoretically attractive feature that it will leave resource allocation unaffected. The practical difficulties of realising this ideal, however, will become apparent in later sections.

Sections 6.2 and 6.3 present a primarily theoretical analysis of the effects of various policy instruments (such as severance taxes, property taxes and income taxes) on the rate of depletion of natural resources, and a discussion of the problems involved in attempting to tax economic rent. Section 6.4 briefly reviews some non-fiscal alternatives such as price regulation. In Sections 6.5 and 6.6 some practical examples are introduced. We have chosen the British system of taxation in the North Sea and the graduated uranium royalty in Saskatchewan as illustrations of the problems involved in designing practical tax measures which can be expected to achieve specified objectives. Section 6.7 reviews some recent developments in the United States with respect to the taxation of energy resources, in particular the reduction in 1975 of the tax allowances available to the oil and gas industries. A few concluding comments make up Section 6.8.

## 6.2 Taxes and the Depletion Rate

This mainly theoretical section concentrates on a number of tax instruments commonly used in the field of energy policy and their impact on the rate of depletion of natural resources. Severance taxes, property taxes, income taxes and capital gains taxes will be investigated in turn.

### (1) *Severance Taxes*

Severance taxes are simply excise taxes placed on the output of extractive industries. Like other excise taxes they may be 'specific', i.e. levied at a certain rate per unit of physical output, or *ad valorem*, i.e. levied on the price of each unit. In the case of oil, for example, the former might also be called a 'barrelage tax' while the latter would include payments of 'royalty', usually a given percentage of the value of oil production at the wellhead.

To understand the ways in which these taxes might affect the depletion rate it is necessary to return to the model developed by Herfindahl and described in Chapter 3. In the simplest case a given known stock of oil extractable at constant marginal cost is available to the community. It was then established that the depletion rate would be determined from the 'fundamental principle' that the 'net price' of the resource must rise at a rate equal to the rate of interest or

$$P_t - C = \lambda(1 + r)^t, \tag{6.1}$$

where  $P_t$  = price of resource at time $t$

$C$ = marginal extraction cost (assumed constant)

$r$ = rate of interest.

Using this very simple apparatus the impact of a specific tax on output was investigated on the assumption that a ceiling price existed at which demand would decline to zero – perhaps as a result of the availability of some close substitute. The conclusion, illustrated in Figure 3.6 (p. 43), was that a specific tax would lower the rate of depletion and extend the time to depletion. It would also lower the present value of the unworked deposits (or royalty[3] as we termed it). This result seems intuitively reasonable. Because the tax is a fixed sum per unit of output it is clear that the present value of the tax paid will be reduced as the output is delayed.

The case is somewhat different, however, for an *ad valorem* tax. Suppose, for example, that a tax is imposed at 20 per cent of the value of sales. According to our model lowering the rate of depletion is less likely to be such an attractive proposition compared with the specific tax case. The basic reason is simply that any delay in tax payment is now to some extent counterbalanced by the fact that future sales will be made at higher prices and that the absolute amount of tax payable on these sales will therefore be higher.

To be more specific, it is known that before the introduction of the tax, equation (6.1) describes an equilibrium time path of prices. After the imposition of an *ad valorem* tax at rate $v$ however it is found that

$$P_t - C - vP_t = \lambda(1 + r)^t - vP_t$$

or

$$\frac{P_t - C - vP_t}{(1 + r)^t} = \lambda - \frac{vP_t}{(1 + r)^t}. \tag{6.2}$$

The present value of the 'net price' (i.e. price net of both extraction costs and taxes) is no longer constant. If the price of the output were to follow the same course as it did prior to the tax then (again from equation (6.1))

$$\frac{\Delta(P_t - C)}{P_t - C} = r$$

or

$$\frac{\Delta P_t}{P_t} = r\left(1 - \frac{C}{P_t}\right). \tag{6.3}$$

From equation (6.3) it is seen that prior to the tax the price of the resource in the market is increasing at a lower rate than the rate of interest. The rate of increase of the market price would in fact tend towards the rate of interest as the (constant) marginal extraction costs became an ever smaller proportion of the price, or would equal the rate of interest if extraction costs were zero. It follows that with *positive* extraction costs the second term on the right-hand side of equation (6.2) must become smaller over time and that hence the value of the whole expression will increase. Resources will become more valuable in present-value terms the longer their extraction is delayed and the existing time path of prices can therefore no longer achieve equilibrium. Prices must adjust to the new equilibrium condition

$$\frac{P_t - C - vP_t}{(1 + r)^t} = \lambda^*,$$

which will involve higher present prices but a slower rate of increase. Theory therefore suggests that severance taxes of both specific and *ad valorem* varieties will reduce depletion rates but that the strength of this force will be less in the case of the *ad valorem* tax. In the limiting case of zero extraction costs the price of the resource prior to the tax must be increasing at the rate of interest and the right-hand side of equation (6.2) will therefore be constant. Thus the same time path of prices will suit the situation both with and without the tax, and in these rather unrealistic conditions the *ad valorem* tax would be entirely neutral with respect to the depletion rate. With $C = 0$ and letting $P_t = \lambda(1 + r)^t$ equation (6.2) could then be written

$$\frac{P_t - vP_t}{(1 + r)^t} = \lambda(1 - v) = \lambda *.$$

The value of the resource is reduced by the rate of tax and no change in depletion policy on the part of resource owners can reduce the burden.

## (2) *Property Taxes*

A tax levied each year on the value of mineral deposits will tend to advance the rate of depletion. As was shown in Chapter 3 the value of deposits in the asset market (the base of the property tax) will be the present value of the future net profit from extracting and selling them. In equilibrium this value will be increasing over time at the rate of interest, thus providing resource owners with the incentive to hold them. Clearly an annual tax on property will greatly reduce this incentive since the longer a deposit is held the greater will be the tax paid on it.

To see in more detail how depletion rates will be affected by a property tax consider the position of the owner of a resource deposit who is deciding whether to hold the deposit for a further period or to extract the resource and sell it in the product market. Suppose that before the property tax is introduced the owner is indifferent between these two prospects as the simple model of Chapter 3 suggests he will be if equilibrium prevails. If $\lambda$ represents the asset value of the resource at time zero and $g$ is its rate of increase then, following the introduction of the tax at rate $p$, a decision to conserve the asset will result in wealth next period of

$$W_1^C = \lambda(1 + g)(1 - p).$$

Extracting the resource and investing the proceeds at the prevailing rate of interest $r$ however will yield

$$W_1^D = \lambda(1 + r).$$

The post-tax equilibrium in which resource owners are once

more indifferent between holding and exploiting deposits will therefore be characterised by the relationship $W_1^C = W_1^D$ or

$$(1 + g) (1 - p) = 1 + r,$$

hence $\quad (1 + g) = \dfrac{1 + r}{1 - p}$ , $\hspace{3cm}$ (6.4)

which for $0 < p < 1$ implies $g > r$. Deposit-holders must be compensated for the property tax by a larger expected rate of increase in the value of their asset. This is equivalent to saying that the rate at which asset-holders discount future net revenue from exploiting their deposits rises with the introduction of the property tax. The result of such an increase in the rate of discount applied to future net revenues is a rise in the depletion rate and a fall in the asset value of the deposits ($\lambda$) compared with the situation prevailing before the tax was introduced.

It is worth noting, however, that the above analysis assumes that the property tax is not integrated into a more general system of wealth taxation. If, for example, all wealth including financial assets as well as physical property were taxed at the same rate there would be no possibility of avoiding tax by switching assets. Such a tax would therefore leave the depletion rate unaffected.[4]

## (3) *Income Taxes*

The incidence of income taxes, especially corporate income taxes, has been a matter of great controversy for many years now and is still far from resolved. Perhaps the simplest approach at the beginning is to assume an income tax is imposed on the natural resource companies alone. It is then fairly straightforward to show, once more using the basic model of Chapter 3, that providing the tax is on *net* income and providing it is a flat-rate tax and not expected to alter in the future, depletion rates will be unaffected.[6] The reason for this lack of 'leverage' is that a 'net price' of the resource rising at the rate of interest will imply a tax take also rising at the rate of interest. Altering the date of production will not therefore change the present value of the tax paid.

Multiplying both sides of equation (6.1) by $(1 - y)$ where $y$ is the rate of income tax we obtain

$$\frac{(P_t - C)(1 - y)}{(1 + r)^t} = \lambda(1 - y). \tag{6.5}$$

A time path of prices which satisfies equation (6.1) will also satisfy equation (6.5) and the present value of the after-tax return remains independent of the time at which production occurs. The tax falls entirely on the resource-owner the value of whose deposits fall from $\lambda$ to $\lambda(1 - y)$.

If the tax is imposed on net business income generally, however, we cannot avoid the question of whether such a tax has a differential impact on the mineral industries *vis-à-vis* the others. Let us suppose that a general tax on business income is *not* shifted forward. The implication is that post-tax rates of return will now be lower than the $r$ per cent previously obtained on productive investments. However, as we have seen, the return to holding mineral deposits remains at $r$ with the depletion rate unaffected in the first instance by the income tax. Clearly this situation cannot continue, and in the long run the expected response would be an increased demand to hold mineral deposits, thereby bidding up their price, with future net revenues from extraction discounted at the now lower prevailing post-tax rate of return. A lower rate of discount then suggests the familiar result that the rate of depletion will be retarded compared with the initial no-tax position.

This result is perhaps counter-intuitive at first sight. We seem to have demonstrated that far from being neutral a business income tax which is not shifted will slow down the rate of depletion. The reason lies in the implied treatment of capital gains. Although the tax is levied on business 'income' we have not included capital gains on the holding of mineral deposits in the tax base and it is this favourable treatment of capital gains which leads to the slower depletion rate.

## (4) *Capital Gains Taxes*

The relationship between the time path of asset prices and the income tax and capital gains tax rates can be illustrated by considering the problem faced by resource owners in deciding whether to extract or conserve their deposits. An equilibrium time path of depletion requires that the owner is indifferent between extracting an extra unit of resource and selling it now, or holding it until the next time period. If the owner decides to exploit the resource immediately, he can sell for a net price $P_0 - C = \lambda$ leaving him $(1 - y)\lambda$ after income tax. By the next period his wealth will have increased to $(1 - y) \lambda(1 + r)$ if invested at the rate of interest, but he will also have to pay income tax on the interest component. Hence the owner's wealth in period one ($W_1^D$) can be written

$$W_1^D = (1 - y) \lambda(1 + r) - yr(1 - y)\lambda$$

or $\qquad W_1^D = (1 - y) \lambda(1 + r - yr).$

If, on the other hand, the resource-owner decides to conserve his deposit until period one it will have grown in value to $(1 - y) \lambda(1 + g)$, where $g$ is the rate of increase in asset prices. Capital gains tax at rate $k$ will be levied on this increase, however, and in these circumstances the owner's wealth ($W_1^C$) can be written

$$W_1^C = (1 - y) \lambda(1 + g - kg).$$

Equilibrium requires therefore that $W_1^D = W_1^C$

or $\qquad 1 + g - kg = 1 + r - yr$

or $\qquad g = \dfrac{r(1 - y)}{1 - k}$ .                                    (6.6)

If $k < y$ it follows that $g < r$ and asset prices rise more slowly than the market rate of interest. This is not unreasonable. If capital gains are treated more favourably than other gains, people will try to achieve income in this favoured form. In the present context this implies holding mineral deposits as assets, i.e. conservation.[7]

## (5) *Tax Allowances*

Some of the major controversies in energy tax policy centre around the impact of tax allowances in the context of corporate income taxes. As we have seen, the objective of the corporate income (or profits) tax is to tax the *net* revenue of an operation, i.e. that revenue remaining after all the costs incurred in earning it have been subtracted. Enormous practical difficulties arise, however, in defining what 'allowable' costs should be in any particular year.[8] For present purposes, however, it is sufficient to concentrate on two major issues – the effect of 'expensing' provisions and 'depletion allowances'.

## (a) 'Expensing'

The traditional treatment of durable assets is that their historic costs should be written off gradually over their working lives. 'Expensing' refers to the practice of treating certain durable assets as if they were current expenses and allowing their cost to be written off against tax immediately. This results in a benefit to the producer in so far as his tax payments are delayed until future periods.[9] When investigating the impact of capital allowances on depletion rates it is obviously necessary to have a 'norm' or standard against which any change can be measured. In his paper on petroleum conservation, McDonald[10] takes as his point of reference the optimal depletion rate prior to the introduction of an income tax and then looks at the combined effects of the tax and expensing provisions. He argues that such a package is neutral with respect to known and already developed deposits. Clearly as far as existing capacity is concerned the related tax allowances will be uninfluenced by changes in the rate of depletion, providing that the allowances are dependent only on the passage of time. Whether immediate write-off or more traditional depreciation procedures are adopted for tax purposes cannot affect the optimal time path of production on already developed resources.

However, an important policy dispute in the United States concerns the likely results of expensing provisions which favour petroleum companies over other firms. Expensing in this debate refers to the ability of petroleum companies to write down 'intangible costs of drilling' (mainly labour costs) immediately, when they should more properly be regarded as an investment necessary to acquire a durable capital asset (i.e. a productive well). This privilege, by increasing the rate of return in the mineral industries relative to that obtainable in other sectors would be expected to induce a flow of resources towards finding and developing petroleum assets until post-tax returns were equalised. In consequence the special tax provisions might result in an over-intensive use of natural resource deposits. In the case of our simple model of depletion with known and fixed stocks of resources, however, the main impact would be on the price of resource deposits with resource owners benefiting from the tax advantages through higher asset values or 'royalties', as we termed them in Chapter 3.

*(b) Depletion Allowances*
A depletion allowance is theoretically no more than the equivalent of a normal depreciation allowance designed to allow for the fact that the extraction of mineral deposits inevitably reduces the total value of those remaining. The mine or well is regarded as a capital asset which must be depreciated as the stock of resource and physical capital is used up. The tax instrument which has given rise to considerable controversy in the United States is the 'percentage depletion deduction'. Until it was substantially withdrawn in 1975 percentage depletion enabled oil and gas producers to deduct 22 per cent of their *gross* income from sales as a depletion allowance. More details of the history of percentage depletion are given in Section 6.8. Here we are merely concerned with the likely impact of this kind of production-related allowance on the rate of depletion.

It is fairly easy to show that the percentage depletion deduction is equivalent in nature to an *ad valorem* sales subsidy. Post-tax income from a unit of production where percentage depletion is available can be written

$$V_\delta = (P_t - C) - y(P_t - C - \delta P_t),$$

where, as before, $P_t$ = price of output, $C$ = marginal cost, $y$ = income tax rate, and $\delta$ = percentage depletion allowance. If, instead of the depletion allowance, a sales subsidy were available, post-tax income could be written

$$V_s = (P_t - C + sP_t) - y(P_t - C),$$

assuming that subsidy receipts are not themselves subject to income tax and where $s$ = subsidy rate.

Setting $V_\delta = V_s$ it is then found that

$$sP_t = y\delta P_t$$

or $\quad s = y\delta.$ (6.7)

An *ad valorem* sales subsidy of $y\delta$ would leave a mining firm in exactly the same position as a percentage depletion allowance of $\delta$. If subsidy receipts are subject to income tax the equation must be modified to

$$s = \frac{y\delta}{1 - y}.$$

With the equivalence of the percentage depletion allowance to a negative *ad valorem* sales tax established, its effect on the rate of depletion can be investigated using the same apparatus as that developed above for severance taxes. It was argued there that an *ad valorem* severance tax would tend to retard the rate of depletion and reduce asset values. By symmetrical reasoning it can be established that a percentage depletion allowance based on the gross value of sales will tend to advance depletion and raise asset values. It is worth pointing out, however, that the strength of this force towards increased present production is not as great as might be expected from considering the impact of a sales subsidy in traditional static analysis. If depletion rates are ultimately related to the 'fundamental principle' and future prices are expected to be higher than current prices the nominal value of future allowances linked to these future prices will also be higher. In the limit, with zero extraction costs and prices rising at the rate of interest, the percentage depletion allowance would be neutral with respect to the depletion rate.

### Limitations of the Simple Model

The analysis above has yielded some fairly definite theoretical predictions but, as is often the case, only at the cost of making some important assumptions. It will be recalled that in the simplest model of Chapter 3 (p. 37) the total resource stock was assumed known, property rights were well-established, the resource was homogeneous, marginal costs were constant, markets were assumed competitive and futures markets worked efficiently. Relaxing these assumptions adds considerably to the complexity of the analysis. The most obvious point is that resource stocks are not known but must be discovered and developed, and that the tax instruments discussed above will have an effect on exploration activity and investment in reserves, and not merely on depletion. 'Expensing' and low capital gains tax rates, for example, will make searching for mineral deposits an attractive proposition as has already been noted, and in conditions of great uncertainty we can no longer deduce that the major beneficiaries of favourable tax provisions will inevitably be the landowners.[12] Further, resource stocks are not fixed in supply but rather are available in varying grades and qualities and at varying extraction costs. As we note in Chapter 2 (p. 24), oil is available in very large quantities from shales and tar sands, but only at higher costs than more conventional sources. Tax measures which raise rates of return in the petroleum industry relative to other industries may therefore have an effect on depletion rates by making the exploitation of hitherto marginal deposits profitable and thus effectively increasing the total stock of the resource economically recoverable. This would apply both at the extensive margin (e.g. rendering the exploitation of shales attractive) and at the intensive margin (e.g. making profitable the installation of equipment in existing wells to improve recovery ratios).

With these reservations in mind the basic results of this section can now be briefly summarised, severance taxes of both specific and *ad valorem* varieties tend to delay depletion with the former being more effective from this point of view. Favourable capital gains provisions relative to the taxation of other forms of income will also tend to encourage conservation. A tax on net income which is not shifted[13] will be neutral with respect to existing deposits, but combined with a depletion allowance based on gross sales will increase the depletion rate. 'Expensing' provisions and other favourable tax allowances may encourage exploration and investment in 'reserves'.

## 6.3   Taxes and Economic Rent

As was argued in Section 6.1 governments have become increasingly concerned to tax the very large economic rents which can arise in minerals industries, more especially of course in the production of crude oil. Before going on to investigate the ability of various tax measures to appropriate this rent it is important to understand how this 'surplus' arises.

In the case of our simple depletion model of Chapter 3 (p. 37) with a fixed stock of a homogeneous resource the analysis approaches most closely the classic view of rent as the return to a factor in inelastic supply. Suppose that a fixed supply of a non-depletable resource existed. Its price would be dependent entirely on its value in production at the margin, and in the absence of maintenance costs the entire price would represent economic rent.[14] Where the stock is depletable, as in the case of energy resources, the analysis is more complicated, as we have seen, but the principle is unaltered. The price of the resource will reflect its marginal value in production, but now this value must include the possibility of using the resource in various time periods and not merely in various uses at a given point in time. An excess of price over marginal extraction costs will represent rent which in equilibrium will rise at the rate of interest. The actual rent received by the producer or owner will therefore depend upon the time at which the resource is exploited, but its present value is independent of this date.

Where costs of production differ, so also will the rents derivable from the mineral deposits. Here the analysis mirrors closely the traditional treatment of differential rents stemming from the varying fertility of land or the more convenient location of markets. As we saw in Chapter 3, higher-cost deposits will yield rents of a lower present value than lower-cost deposits. This may be because nominal rents are lower when exploitation occurs, but a more important factor is that the higher-cost deposits will be exploited later (see Figure 3.7, p. 46).

Casual observation suggests, however, that our basic theory is deficient. Low-cost sources of oil from the Middle East are produced along with oil from inhospitable regions of Alaska and offshore deposits in the Gulf of Mexico or the North Sea. It seems that other considerations influence resource exploitation decisions than the profit-maximising behaviour of competitive oil companies. Two important possibilities are the desire for security and the existence of monopoly.

Importing oil from low-cost regions has the obvious political disadvantage that such dependence could be used as a bargaining counter in negotiations on quite unrelated topics – the recent Middle East oil embargo is a clear example. The problem is inevitably accentuated if the low-cost areas are very concentrated geographically and organised in a tightly knit cartel such as OPEC (the Organisation of Petroleum Exporting Countries). From the point of view of any individual country the financial costs of obtaining oil are not therefore the only costs to consider and there may exist an incentive to place a tariff or other restriction on imports in order to encourage domestic production of higher-cost reserves. This will have the added effect of increasing the rents derivable from the more favourable domestic sources.

Monopoly control of all oil deposits would not be expected to result in these deposits being exploited in an inefficient sequence. They would simply be exploited more slowly with higher present prices. In present conditions, however, OPEC does not control all oil deposits, merely the least-cost ones. By setting prices higher than a competitive market would produce, OPEC thereby open up opportunities for hitherto rather marginal deposits to be exploited profitably. It should be noted, however, that higher prices *per se* will not necessarily lead to exploitation of the higher-cost resources. If the rents from OPEC oil were expected to continue to rise in the future at the rate of discount used by the owners of high-cost reserves then the analysis of Chapter 3 indicates that the latter will still find it expedient to delay production. If on the other hand the expectation was that the OPEC price was 'too high' in the sense of being above a sustainable equilibrium time path, and that it was unlikely to rise much further for several years, there would be a clear incentive to deplete high-cost reserves as quickly as possible.

Thus, for the purposes of analysing the effects of various

taxes on rents it is useful to distinguish between a framework in the traditions of the simple depletion model of Chapter 3 and a framework which regards rent as primarily the outcome of tariff protection or monopoly pricing.

## Rents in the Depletion Model

Although only passing reference was made in the last section to the effects of the tax measures investigated on asset values or 'royalties', it is evident that these indicate the extent to which taxes are borne by resource-owners (i.e. the extent to which they reduce the present value of the rents). Briefly the results indicated that severance taxes would reduce rents and in the limiting case of an *ad valorem* tax with zero extraction costs would be fully paid out of rents. Similarly, taxes on wealth and net income (in the absence of special expensing or depletion provisions) will fall entirely on the owners of resource deposits. Property taxes are also predicted to result in a drop in asset values, but the fact that this tax is not neutral with respect to the depletion rate even in the simplest of models suggests that the final incidence is more complicated.[15]

## Rents in a Stable Prices Model

In this section the price of the extracted resource is assumed to be set by a cartel outside the control of the producing companies. This price is expected to remain constant and there is therefore a strong incentive to deplete quickly. Faster depletion is costly, however, requiring more wells and lower ultimate recovery and companies therefore select a rate of depletion which maximises the net present value of their assets. It is further assumed, not unrealistically, that the quality of deposits within the boundaries of the country concerned vary widely and that higher outputs per unit of time can only be obtained by resorting to more costly sources of supply. More definite examples of cost variations in the North Sea are given in Section 6.5. In these circumstances the supply curve of output will be upward-sloping as represented in Figure 6.1.

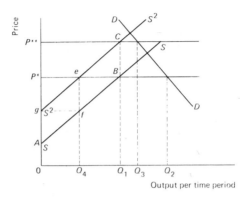

FIGURE 6.1

With the demand curve $DD$ and the import price fixed at $P^*$ output will be $0Q_1$ and imports will amount to $Q_1 Q_2$. Total rent from production per time period will be given by the area $AP^*B$. Standard supply and demand analysis can now be used to demonstrate the effect of a specific severance tax. Following Kemp[16] we first assume that the tax is placed both on domestic and imported sources. The supply curve shifts as a result to $S^2 S^2$ and the import price to $P^{**}$. It is evident that such a measure does nothing to capture the economic rent. The main effect is to reduce total demand and imports by the same amount. A tax on domestic supplies only, however, is different. In this case the market price remains fixed at $P^*$ and the revenue raised ($P^*efg$) comes entirely from domestic producers. Imports rise from $Q_1 Q_2$ to $Q_4 Q_2$. A severance tax will fall on rent therefore, providing that imports are not taxed.[17]

As with the model of Section 6.2 a profits tax in the stable prices case would fall substantially on rent since the assumption of a fixed import price ensures that the tax cannot be shifted forwards. In principle such a tax should prove a much more efficient 'rent extracter' than a severance tax and would avoid pricing the most costly sources out of the market. In practice it is doubtful whether profits taxes exhibit this degree of neutrality.[18] Allowances for capital depreciation, for example, will often favour certain projects over others. A more direct way of taxing net present values would be to do so via a property tax. The analysis here is little different from that of Section 6.2. The tax would fall on resource rents, but there would be additional effects associated with speeding up the rate of depletion.[19]

Criticisms of these tax measures can be made on several grounds. The severance tax on domestic production alone has associated efficiency losses in that the resulting increase in imports could have been produced from domestic reserves at lower resource cost; the property tax hastens depletion rates; while both taxes will have adverse effects on company liquidity at the early stages of development when they could become payable before positive net revenues are achieved.[20] Unless these effects are desired for quite separate reasons of policy they obviously suggest the possibility of improved means of taxing rent. In principle after all, as was pointed out in Section 6.1, the advantage of taxing rent is that it can be accomplished without inducing efficiency losses through the reallocation of resources. Further, a major criticism of all the tax measures so far discussed is that they are capable of yielding to the exchequer only a proportion of the total rents available. A profits tax levied at a flat rate of 50 per cent, for example, would still leave 50 per cent of the rents in the hands of the companies and would therefore still imply the existence of higher post-tax rates of return on low-cost deposits relative to high-cost deposits.[21] These considerations have led a number of economists to consider alternative ways of taxing rent.

## Progressive Rent Taxes

The first possibility to consider is that a profits tax could be levied at increasing marginal rates instead of being levied at a flat rate. Unfortunately this does not solve the problem. It is quite possible for higher *absolute* profit levels to be achieved on high-cost reserves requiring very large quantities of capital investment if the volume of these reserves is great. Smaller deposits, on the other hand, yielding smaller flows of net revenue might nevertheless produce much larger profits per unit of production, or per unit of capital invested. Thus there is no assurance that a profits tax levied at increasing marginal rates would achieve the objective of appropriating economic rent.

Economists have not, however, been slow to suggest refinements. If profits per unit of production vary a great deal it has been suggested that the rate of tax could be determined with reference to this ratio. Thus a deposit with a higher ratio of profits to output in any year would be required to pay a higher rate of tax on the profits. Alternatively, Kemp proposes to link the tax rate to the ratio of profits to total capital expenditure.[22] In either case the objective is to tax more heavily those projects yielding a higher rate of return, with profits per unit of output or profits per unit of capital expenditure as a rough guide.

The extreme difficulty of devising a scheme able to cream off economic rents from oilfields of greatly differing sizes and capital requirements in the North Sea without simultaneously jeopardising the development of marginal fields has led some researchers to the conclusion that the effective rate of tax might be better decided *ex post* by setting an 'acceptable' rate of return and then 'ensuring that this return is met on each field'.[23] In fact this idea amounts to a special case of a more detailed proposal by Garnaut and Clunies Ross[24] for a resource rent tax in which the tax rate would vary with the rate of return. Their proposal, put forward as a device for developing countries to benefit from high natural resource prices, is that profits should be taxed after a 'threshold internal rate of return' has been achieved.

Consider, for example, a project involving a base year outlay $(K_0)$ and then a stream of net revenue $(R_i)$ over $n$ years. Its net present value $(V)$ will be

$$V = -K_0 + \frac{R_1}{1+r} + \frac{R_2}{(1+r)^2} + \frac{R_3}{(1+r)^3} + \cdots + \frac{R_n}{(1+r)^n},$$

where $r$ is the rate of discount applied by the company undertaking the project. Now suppose that the Government announces its threshold internal rate of return $p_1$. Further suppose that, when discounted at rate $p_1$, we have

$$K_0 = \frac{R_1}{1+p_1} + \frac{R_2}{(1+p_1)^2}.$$

In this situation net revenue in the first two years of the project would suffice to yield the required internal rate of return $p_1$ on the capital investment. The objective would then be to leave these revenues untaxed. Tax (say at rate $t_1$) would become payable only on the later revenues $R_3, R_4, \ldots, R_n$, i.e. those causing the rate of return to exceed $p_1$. The tax could be made progressive by declaring in advance several threshold rates of return along with the tax rates applicable to them. It might happen, for example, that at discount rate $p_2 > p_1$

$$K_0 = \frac{R_1}{1 + p_2} + \frac{R_2}{(1 + p_2)^2} + \frac{R_3}{(1 + p_2)^3} .$$

Revenues $R_4, R_5, \ldots, R_n$, i.e. those causing the rate of return to exceed $p_2$ could then be taxed at $t_2$. After repeating this exercise at each threshold rate the final result would be to leave $R_1$ and $R_2$ tax free, to subject $R_3$ to a tax of $t_1$, and to tax $R_4$ and future net revenues at $t_1 + t_2$.[25]

There are unfortunately major problems associated with most schemes for progressive rent taxes. As the proponents of the various ideas all recognise, setting an 'acceptable' rate of return and taxing excess profits at very high rates removes the incentive to adopt economically efficient techniques. This can be seen most clearly in the case of a 100 per cent tax on profits above a given rate of return. If this rate is set below the company's discount rate then no resource deposits will be developed. If, on the other hand, it is set higher than this the company will have an incentive to avoid tax and appropriate the available rent in the form of an 'acceptable' rate of return on totally unproductive investment. For similar reasons Sumner[26] has recently shown that a progressive rent tax of the variety described earlier will not usually be neutral with respect to the rate of depletion. Assuming for a moment that the Government was vigilant in preventing totally unproductive investment, it would still be open to the companies to invest extra capital in advancing production and increasing the rate of output. It would appear therefore that progressive rent taxes will often require extensive government surveillance of mineral company policy and that although specifically designed to tax rent they lack the neutrality which would characterise the ideal tax instrument.[27]

## Competitive Bidding

At first sight it might be supposed that the design of sophisticated tax measures is unnecessary because simpler and more direct measures for appropriating rent exist. Theoretically at least, all the rent could be obtained by a process of competitive bidding. Firms would be required to obtain a lease from a public licensing authority before they were permitted to develop a resource. This lease would be granted to the firm offering the largest 'lease bonus', i.e. a lump-sum payment to the public authority. The maximum sum competitive firms would be willing to pay would be that which reduced the present value of the project to zero when evaluated at their own discount rate. In a competitive auction this sum would be obtained and the full present value of the economic rents from the project would immediately accrue to the exchequer. The system would have the added advantages that it would not discourage the development of marginal high-cost deposits (these would simply yield lower or zero lease bonuses), and that control over mineral deposits would be in the hands of the most efficient firms (those able to offer the highest bids). The lump-sum payment would clearly not affect the rate at which the firm found it profitable to deplete the resource.

It is evident that the competitive-bidding approach has great theoretical attractions but its operation in practice will depend on how closely conditions approach the implicit assumptions underlying the theory. For an auction system to be perfectly efficient the usual assumptions of perfectly competitive analysis are required to hold, most notably perfect knowledge, the absence of uncertainty and no collusive behaviour. However, to rule out a scheme on the grounds that these conditions are not fulfilled would be to compare a working possibility with a textbook ideal, when the real choice is between various institutional arrangements all with associated practical difficulties. There exists considerable disagreement on the relative effectiveness of auction systems. Critics point to several problems.

(*a*) Competitive bidding is a possibility only if the resources

concerned are located in public land. If rents can be approp-
riated by landowners alternative fiscal measures will be
required to return them to the exchequer.[28] In practical
terms this is not such a severe problem as it might at first
appear, especially in the case of petroleum reserves. Offshore
oil reserves, which are becoming increasingly important, are
under the control of the littoral states,[29] the extensive oil
discoveries in Alaska are on federal land, while shale deposits
in Colorado, Utah and Wyoming are mainly Federally
owned.[30] Similar comments apply to natural gas resources in
Canada where large quantities have been discovered in the
Mackenzie Delta and the Arctic Islands.

(*b*) Collusion on the part of companies bidding for leases
might clearly reduce the rent which the state could obtain. A
vigorous restrictive practices policy would therefore have to
be an accompanying feature of an auction system. Whether
petroleum companies are sufficiently competitive is ultimately
an empirical question concerning which there is much disagreement.[31]

(*c*) Lease bonuses are payable immediately and therefore
put considerable strains on a firm's 'cash flow'. Fears have
been expressed[32] that this might inhibit exploration and
development. This criticism is clearly identical to that levelled
at property taxes and *ad valorem* sales taxes which we
encountered earlier. As before it depends for its validity on
imperfectly operating capital markets. Supporters of competitive bidding[33] make two basic points in reply.

(i) Companies are observed to be willing to pay large sums
for exploration, and development rights, for example in the
Gulf of Mexico and Alaska, without any obvious retarding
influence on the rate of development. Indeed it is sometimes
argued that the greater the lease bonuses paid the larger will
be the incentive to avoid unnecessary delay.

(ii) Immediate cash bonuses are not an inevitable part of a
bidding system. Bids might be accepted in the form of a
royalty on future production, a device which would have the
advantage of enabling smaller firms to tender thereby increasing the competitiveness of the auction. It would also, of
course, have the disadvantages associated with barrellage or
*ad valorem* sales taxes analysed above.

(*d*) Probably the most telling criticism of competitive bidding is that it is inappropriate in conditions of extreme risk or uncertainty. When exploratory drilling is about to take place in a new location, oil companies may have only the haziest idea of the probability of finding a deposit, much less of its size and commercial value. In these circumstances bonus bids are unlikely to bear much relation to the final rents (if any) which accrue to the firm.[34] On the other hand it could be argued that the Government itself could improve information by undertaking or subsidising geological surveys and that in any case it does not have to license all areas at once but can issue them gradually as new knowledge becomes available. The effects of such knowledge can be startling, as Table 6.1 below on lease bonuses paid on the North Slope of Alaska demonstrates. Bonuses paid per acre rose from $39 to $2182 between the 1967 and 1969 sales.

TABLE 6.1

*Results of North Slope Lease Sales*

| Date of sale | Acres leased | $/acre | Bonuses paid $ |
|---|---|---|---|
| 9 Dec 1964 | 466,180 | 9 | 4,376,523 |
| 15 July 1965 | 403,000 | 15 | 6,145,473 |
| 24 Jan 1967 | 37,662 | 39 | 1,469,645 |
| 10 Sep 1969 | 412,548 | 2182 | 900,218,590 |

SOURCE:[35] Gregg K. Erickson, 'Alaska's Petroleum Leasing Policy', *Alaska Review of Business and Economic Conditions*, vol. VII, no. 3 (July 1970) p. 4.

Further, other mechanisms exist within the structure of licensing arrangements to recoup at least a portion of any unexpectedly large rents obtained. Most licensing systems include 'surrender provisions' by which, after a lapse of time, a proportion of the licensed area is returned to the Government and may be relicensed. The proportion is usually 50 per cent, but the detailed provisions can vary a great deal from being entirely at the discretion of the licensee to Alberta's checker-board system by which the retained portion must be arranged in blocks which touch each other only at the corners or are separated by at least one mile. Surrender provisions may have the additional effect of speeding up exploration by giving licensees an incentive to discover the most valuable portions of the licensed area before the surrender date.

Even assuming sufficient knowledge of geological and other conditions to permit a fairly objective assessment of the probabilities of various outcomes, the existence of risk still raises difficulties for a competitive auction. Basically the problem is one of risk-spreading or risk-pooling. If probabilities of success or failure really are objectively known, a sufficiently diverse portfolio of 'prospects' should enable a firm to offer as a lease-bonus the expected value of the rents derivable from each project. In some areas it would lose perhaps the whole of the bonus if holes proved to be dry; in others it might strike a rich deposit. The overall result, however, would be that the state would receive the present value of all the future rents. This happy result requires, however, not only that probabilities be known, but also that firms are large enough to undertake a sufficient number of ventures to spread their risks widely, or, if not, are able to join with other firms in developing a similarly wide range of options. Clearly both the existence of very large firms and joint bidding arrangements could pose problems for maintaining a competitive auction.[36]

Supposing now that there are institutional limits to the amount of risk-pooling possible. Each firm cannot avoid bearing some risk and must be compensated for doing so if exploration is to take place. It is evident that the bonus they would be willing to offer would be lower than those in the risk-pooling case and the Government will receive less than the expected present value of future rents (or put differently it will receive the expected present value of future rents evaluated at the firm's new 'risky' discount rate). In these circumstances both Government and firm could benefit from a rather different form of 'risk-sharing'. By accepting lower lease-bonuses but levying higher taxes later in the event of a successful outcome, the Government would reduce the risk to the company by lowering the probability of either very high or disasterously low returns. Such a measure could increase the present value of the Government's take and make marginal prospects more attractive to the firm once more.[37] Thus it appears that competitive bidding in the form of cash bonuses may be inappropriate in the presence of risk since any lump-sum payment made before the beginning of a project can only increase the chance of receiving very low or even negative returns.

(*e*) A criticism related to point (*d*) above is that any unanticipated rise in the resource price and hence in resource rents would remain untaxed under a competitive bidding system. Apart from the effect of 'surrender provisions' already described there is clearly some force in this argument. The most celebrated example is, of course, the enormous increase by OPEC in the posted price of oil at the outbreak of the Arab–Israeli war in October 1973 and January 1974. It should be remembered, however, that a lower than expected rate of increase in resource prices would have precisely the opposite effect, the Government receiving in such a case a larger sum than the total rents eventually gained by the firms.

It is evident that no single scheme is appropriate for all possible cases as a device for taxing rent. Income taxes and severance taxes will fail to appropriate all rents and may discourage the development of marginal resources. Most schemes for a progressive rent tax have the side-effects of encouraging waste and altering the rate of depletion. Competitive bidding is attractive theoretically but may be inferior to alternatives in conditions of extreme uncertainty, monopoly practices or imperfect capital markets. These possibilities are not mutually exclusive, however, and elements of competitive bidding can be retained in conjunction with other measures. Firms could be invited to bid for leases, for example, on the understanding that successful projects will be required to pay an 'excess profits tax' or a royalty on production at a specified rate.[38] The final instrument or combination of instruments chosen will therefore depend on the Government's assessment of conditions prevailing in the relevant markets (e.g. with respect to uncertainty), its estimates of the likely quantitative importance of some of the qualitative results derived above (e.g. on the question of the discouragement to the development of marginal deposits), and the importance of other objectives that it may be pursuing in addition to the simple appropriation of economic rent.

## 6.4 Non-Fiscal Instruments

Before considering some of the attempts which have been made at overcoming these problems it is worth casting a sidelong glance at some of the alternative non-fiscal instruments at a government's disposal.[39] We shall look in turn at price regulation, monopoly purchasing agencies and 'participation agreements'.

### (a) *Price Regulation*

If producers are expected to receive large rents from minerals exploitation one possible approach is to regulate the price. Elementary economic theory is enough to predict the consequences. If the regulated price is set below the market price some of the rent will be passed on to consumers, but the lower price will also increase the quantity demanded and lower the quantity supplied thereby creating a 'shortage'. In Figure 6.2 a reduction in price from the equilibrium level of $P_E$ to a regulated level $P_R$ enables consumers to gain the rent $P_R abP_E$, but also creates excess demand of $q_1 q_2$.

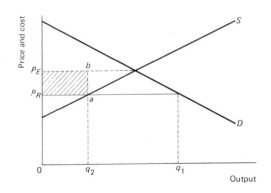

FIGURE 6.2

In practice these effects may take some time to work themselves out. Energy resources tend to be in rather inelastic supply in the short run, while demand for a particular resource may also be inelastic in the short run since domestic appliances or industrial plant can be adapted to take different fuels only to a limited degree. Variations in the rate of output are very costly in the case of oil after production from a particular field has commenced, while in the case of natural gas the necessity of building pipelines to take the gas to the consumer implies that it is sold under long-term contracts committing the producer to maintaining a certain rate of flow to a given pipeline.

Experience in the United States suggests, however, that price controls can have very serious effects in the long term (see Chapter 9, Section 9.4). In 1960 the Federal Power Commission began imposing ceiling prices on producers of natural gas selling in the interstate market. By the early 1970s shortages had begun to develop, showing themselves in the form of increasing waiting lists of customers, and a rising incidence of supply interruptions where contracts permitted.[40] The ratio of proven reserves to production declined substantially as producers failed to replace their 'inventories'. Inevitably discussion was stimulated on the question of how far price regulation was responsible for the shortage, and a number of models developed in the United States tend to verify its importance.[41]

### (b) *Monopoly Purchasing*

By stipulating that mineral supplies may be sold only to a state agency this agency will be able to reduce the price at which it agrees to purchase until it relieves the producer of all rent, leaving the latter with a normal rate of return. In the United Kingdom, for example, natural gas must be offered to the British Gas Corporation (unless it is to be used for non-fuel purposes such as a petrochemical feedstock). The system differs from simple price regulation in that the price the final consumer pays is dependent upon the policy of the monopoly buyer (or monopsonist). If the Gas Corporation charges market prices for its gas, any rent it captures from the producers will appear as profit in its accounts. On the other hand if it sells at lower prices some or all of the rent will accrue to consumers and the effects would be expected to be rather similar to those of price regulation.

How far exploration for new deposits has been affected by the purchasing policy of the British Gas Corporation in the particular case of the North Sea is difficult to judge. Professor Dam,[42] for example, argues that the prices paid in the late 1960s probably did have a substantial disincentive effect although conclusive evidence is difficult to obtain since observed changes in exploration activity, most notably a shift to more northerly waters where gas is more likely to be found in association with oil, are compatible with purely geological explanations. Certainly there has been none of the obvious symptoms of shortage mentioned in subsection (*a*).

Whatever may have been the case in this particular example, the principal point at issue is that a departure from market prices whether by regulation or monopsony power inhibits the eventual prices agreed in one of their main functions — allocating resources to future development. The return to a mineral resource producer is not a 'pure' rent in the sense of being a return to a factor in totally inelastic supply. Resources are used up and require replacement, and it is prices, both present and those expected in the future, which, as we have seen, determine the rate at which deposits are discovered and developed. It is for this reason that some analysts prefer the term 'quasi-rents' when applied to the returns from minerals exploitation. The danger is that, by fixing prices at too low a level relative to those expected to prevail in the future, or by simply adding an extra element of uncertainty to the process of price-formation, a regulatory agency or monopsonist will adversely effect the incentive to develop new and possibly higher-cost deposits.

## (*c*) *Participation*

One straightforward way for the state to receive the rents from the development of mineral resources is for the state to provide the capital and undertake the development. The practical objection to such a proposal is that the technical expertise is usually in the hands of private corporations and that to rely on state enterprise would lead to delay. A convenient compromise is the 'participation agreement' in which the state agrees to provide a proportion of the capital in conjunction with private corporations and then receives the same proportion of the profits. The ability to extract rent by this means thus depends on the extent of the participation.

Since participation involves an agreement between Government (or their agents) and private companies their precise form can obviously be very varied. In the North Sea, however, the system used in both Norway and the United Kingdom is the so-called 'carried interest' system. A company is licensed to explore in certain areas on the understanding that, in the event of a commercial discovery, the Government may take an interest (say 51 per cent) in the project. Until that time the Government's interest is 'carried' by the oil company concerned. In both the United Kingdom and Norway public corporations have been set up to oversee the state's interest in offshore projects (Statoil in Norway and the British National Oil Corporation (B.N.O.C.) in the United Kingdom). For the purpose of ensuring that the state has a financial interest in offshore ventures these companies would require only managerial and financial expertise, but in fact both Statoil and B.N.O.C. have wide powers to enter the fields of exploration, production, refining and marketing.[43]

Whether participation has any disincentive effect on exploration and development by private companies is impossible to judge *a priori*. In principle, a share by a partner in the financial costs should not affect the rate of return on the capital invested and if the Government also shared the 'dry hole' costs there could even be an incentive effect via the reduction in risk and a wider spread of activity. However, the 'carried interest' system is specifically designed to relieve the state from contributing to unproductive ventures. Further, a 51 per cent share by a state agency gives the latter wide control over such matters as depletion policy which might conceivably be of great significance in determining the profitability of any given project. It is this factor of control which is probably the most unpalatable aspect of participation from the point of view of the private companies. On the other hand they may not have to worry. As Professor Dam perceptively comments: 'the state oil company may come to see its interests more nearly congruent with the private companies than with the Government'.[44]


## 6.5   U.K. Taxation in the North Sea

Exploration of the continental shelf under the North Sea was precipitated by the discovery in 1959 of one of the largest gas fields in the world at Slochteren in the Netherlands. Activity was concentrated initially in the southern basin and the first discovery in the British sector occurred in 1965. The major advances took place in 1966, however, with the discovery of the Leman Bank, Indefatigable and Hewett fields. By the end of 1976 total proven gas reserves in the British sector amounted to 809,000 million cubic metres. Production from the fields in 1976 is recorded in Table 6.2.

In the late 1960s interest turned to more northerly waters following the discovery of the Ekofisk oilfield in the Norwegian sector in 1969. By November 1970 B.P. had announced the Forties field off the east coast of Scotland, and the next few years saw equally large discoveries still further north in the East Shetland Basin – most notably the Brent and Ninian fields. Proven reserves in 1977 are estimated to

TABLE 6.2

*Gasfields in production, British Sector of North Sea*

| Field name | Discovery date | Initial production date | Production in 1976 (million cubic metres) |
|---|---|---|---|
| Leman Bank | April 1966 | Aug 1968 | 15,920 |
| Hewett | Oct 1966 | July 1969 | 8,190 |
| Indefatigable | June 1966 | Oct 1971 | 6,560 |
| Viking | May 1968 | July 1972 | 6,180 |
| West Sole | Oct 1965 | Mar 1967 | 2,040 |
| Rough | May 1968 | Oct 1965 | 520 |
| Frigg (U.K.) | May 1972 | Sep 1977 | – |

SOURCE: *Energy*, Central Office of Information Reference Pamphlet 124 (London: H.M.S.O., 1977) table 5.

be 1380 million tonnes although total reserves may prove to be substantially higher.[45] Estimated peak production from the fields presently under development is given in Table 6.3.

For our present purposes three important points require emphasis. First, exploring for petroleum in the North Sea is an activity subject to great uncertainty. The probability of discovering a commercially viable deposit at any given location is exceedingly difficult to judge[45] although experience and increasing knowledge of the geological structure of the North Sea obviously help. Second, extracting oil and gas from beneath the North Sea is technologically very demanding, thus adding considerably to the uncertainties surrounding development costs. Third, fields can vary considerably both in terms of reserves and the capital requirements for development. Fields like Ninian and Brent, for example, are 'front-loaded', that is, they require very large capital outlays at the beginning while revenue from production may not commence until the fourth year, building to a plateau in perhaps the sixth. The requirement for large production platforms and

TABLE 6.3

| Offshore oilfields | Peak production* |
| --- | --- |
| | (million tonnes/year) |
| Forties | 24.0 |
| Auk | 2.5 |
| Argyll | 1.1 |
| Beryl | 4.0 |
| Brent | 23.0 |
| Piper | 12.0 |
| Montrose | 2.4 |
| Ninian | 16.5 |
| Thistle | 10.1 |
| Claymore | 8.5 |
| Dunlin | 7.5 |
| Statfjord (United Kingdom) | 4.2 |
| Cormorant | 3.0 |
| Heather | 2.5 |

* As estimated by operators.

SOURCE: *Energy*, Central Office of Information Reference Pamphlet 124 (London: H.M.S.O., 1977) table 3.

long submarine pipelines to transport the oil to shore[46] implies the commitment of massive funds at an early stage, funds which when discounted at a positive rate appear even more significant in relation to the future net revenues. The net present value of projects such as these is understandably sensitive to changes in expected future price levels and to the rate of discount applied. In contrast, fields such as Auk and Argyll are small by comparison but are developed more quickly with low capital costs per unit of peak production and higher operating costs as a result of using offshore loading into tankers. Their shorter lives and the relatively fast attainment of positive net revenues make these fields far less sensitive to price and interest-rate changes.[47]

These conditions of uncertainty and of considerable variation in the financial outlays required for development make it extremely difficult to devise a tax regime which could be expected to achieve all the objectives that governments appear to have had in mind. The major objectives, according to official statements, are as follows:

(*a*) 'To secure that exploration continues as fast as reasonably practicable.'
(*b*) 'The attainment as early as practicable of net self-sufficiency in oil' for balance-of-payments and security reasons.
(*c*) To avoid any discouragement to North Sea investment[48] but
(*d*) To ensure that the oil companies do not 'reap enormous and uncovenanted profits on their investment'.[50]
(*e*) To favour British interests and to secure for British industry 'a greater share . . . of the £1200 million-a-year onshore/offshore supplies market'.[51]

The instruments at the Governments' disposal for achieving these objectives include licensing, participation, royalties, the petroleum revenue tax and the corporation tax.

## The U.K. Licensing System

For the purposes of licensing the U.K. sector of the North Sea is divided into blocks. Each block is about 100 square miles in area. With the exception of 16 blocks awarded by competitive tender during the fourth round of licensing as an 'experiment' all licences have been issued by ministerial discretion. Thus, in the United Kingdom, licensing has not been used as a device for appropriating economic rent, but has been viewed much more as an instrument of control. Payments are purely nominal – an application fee of £1250 and some additional periodic payments related to the area licensed.[52] In addition a royalty of 12½ per cent of the value of any future production is imposed. Receipt of a licence depends on the potential licensee agreeing to a number of conditions.[53]

(a) The British National Oil Corporation must be a co-licensee with a 51 per cent equity interest in each licence.

(b) All licensees and the Secretary of State must agree on a 'work programme'.

(c) An applicant must also satisfy the Secretary of State with respect to no less than 14 other 'criteria' such as the applicant's past performance 'in providing full and fair opportunity to U.K. industry to compete for orders of goods and services.'[54]

Licensing policy has therefore concentrated on encouraging exploration by insisting on work programmes, on protecting British interests both in the granting of the licences themselves and in encouraging the offshore supplies industry, and, since the fifth round, on enforcing state participation. As we have seen, state participation is a means of securing some of the rent accruing from oil production, but it still leaves the private companies with a share according to the extent of the participation. The 12½ per cent royalty on the gross value of production is aimed at this share, but, as the discussion of Section 6.3 demonstrated, such a tax may adversely affect marginal fields. For this reason the Secretary of State has the power to refund royalties 'for the purpose of facilitating or maintaining the development of the petroleum resources of the United Kingdom'.[55] As a more explicit and carefully tailored means of taxing excess profits the Petroleum Revenue Tax was introduced in 1975.[56]

## The Petroleum Revenue Tax

Petroleum Revenue Tax (P.R.T.) is levied at a rate of 45 per cent on the assessable profit from each oilfield. It is levied prior to corporation tax. Assessable profit is the gross market value of oil sales minus a number of allowances for royalty and periodic licence payments, operating costs and capital costs. Interest payments are not allowable. Additional allowances may be claimed for abortive exploration expenditure or for a loss on an abandoned field (these being an exception to the general rule that the tax should be applied on a field-by-field basis). Losses from previous years may be allowed in computing assessable profits in future years. Three special provisions are particularly important:

(1) In calculating assessable profits companies may immediately deduct all capital expenditure incurred in discovering and developing a field together with an additional 75 per cent of this amount (the so-called 'uplift' provision).
(2) An 'oil allowance' is available equivalent to the value of one million long tons per year on each field. The total oil allowance claimed for each field cannot exceed ten million long tons over the life of the field.
(3) An annual limit is placed on the amount of tax payable – the 'safeguard' and 'tapering' provisions. In particular, if an 'adjusted profit' figure falls below 30 per cent of accumulated capital expenditure, P.R.T. liability declines to zero. If adjusted profit exceeds 30 per cent of capital expenditure P.R.T. liability must not be greater than 80 per cent of this excess. Adjusted profit is gross revenue minus royalties and operating costs.

As already noted, the operating companies must also pay corporation tax on their profits. Profits in this context are net of P.R.T., i.e. for corporation tax purposes the P.R.T. is treated as an allowable cost. North Sea activities are treated rather differently from others, however, in that they are subject to the 'ring fence'. This 'ring fence' prevents a company from using losses or allowances from other activities to set against profits from its oil interests in the North Sea.

However, any losses on North Sea operations can be set against profits elsewhere.

The P.R.T., corporation tax and 12½ per cent royalty payment between them may be considered the 'fiscal package' affecting petroleum production in the United Kingdom. Although only empirical work can demonstrate the impact of the whole package on profitability and production a few observations are sufficient to indicate the likely results.[57]

(*a*) The P.R.T. is levied at a flat rate, a fact which would be expected to limit its success as a rent-extracting instrument. Any progressivity in the 'take' from fields of varying profitability is possible only via the structure of 'allowances' — rather as a flat-rate personal income tax can be made progressive by tax-free allowances.

(*b*) As a means of helping marginal fields the three major provisions concerned are arguably slightly off-target.

(i) The 75 per cent capital uplift is a help to all fields, especially the most capital intensive, but not *necessarily* the least profitable. There is no doubt, however, that the ability to write off 175 per cent of capital expenditure before paying P.R.T. is of great assistance to large front-loaded projects which might otherwise be more sensitive to price and interest-rate changes. This illustrates in stark form the dilemma facing a Government wishing to tax resource 'rents' in conditions of uncertainty. In order to be *confident* that a project will not be discouraged it may be necessary to offer allowances which make it *probable* that large amounts of rent will remain in the hands of the private companies. A progressive rent tax of the type described in Section 6.6 would be required to overcome this dilemma.

(ii) The 'oil allowance' is open to a similar criticism. The limitation of this allowance to one million long tonnes per year implies that it is aimed at helping small fields relative to large ones. However, small fields can be exceptionally profitable as the experience of Auk and Argyll indicate.[58]

(iii) The 'safeguard' and 'tapering' provisions are clearly intended to assist projects on the margin of profitability and may be seen as a crude application of Kemp's idea that the tax rate should be related to the ratio of profits to accumulated capital expenditure (see Section 6.3 above). However, whether the provisions succeed in their objective depends a great deal on whether the companies and the Government agree on what is 'marginal'. According to estimates by Kemp in 1976, removal of the 'safeguard' and 'tapering' provisions would make virtually no difference to the net present values of existing fields given his assumptions about future prices and costs.[59] The implication is presumably that the Government felt that no marginal fields existed in 1976.

Section 6.5 may be summarised as follows. The principal instruments of U.K. Government policy in the North Sea are licensing and state participation through the B.N.O.C. Discretionary licensing inevitably results in rents passing to the oil companies, and the P.R.T. is the instrument intended to reclaim them. The desire to avoid any discouragement to the development of marginal fields and the fact that it is levied at a flat rate probably reduce its effectiveness in achieving this objective.

## 6.6   Uranium Royalties in Saskatchewan

As an example of an alternative system of taxing natural resource companies, and as a reminder that petroleum is not the only important energy resource, it is instructive to consider the scheme introduced in Saskatchewan in 1976. The rising price of oil in the early 1970s led to a sympathetic rise in the price of possible substitutes such as uranium, and the Government of Saskatchewan, a province containing some rich deposits, was placed in a position rather similar to that of the U.K. Government contemplating the development of North Sea oil. As the statement made to the Legislative Assembly of Saskatchewan[60] makes clear, the objectives were 'to ensure that a fair share of the excess profits from uranium minerals is captured by the province as owners of the resource' and 'to leave marginal production decisions as unaffected as possible'.

The scheme finally adopted[61] provides us with a practical example of a progressive rent tax, albeit not of the theoretical purity of those suggested by Garnaut and Clunies Ross or by Sumner (see Section 6.3 above). It contains elements of several of the schemes discussed. The system can be briefly described as follows:

(1) Just as the P.R.T. applies to a particular oilfield, so the uranium royalty system applies to a single mine complex.

(2) A basic royalty of 3 per cent of the gross value of sales is levied each year.

(3) A graduated royalty is payable depending on the ratio of operating profit to capital investment – none below 15 per cent, a 15 per cent tax on profits between 25 per cent and 45 per cent of investment, and a 50 per cent tax on profits above 45 per cent of investment.[62] This aspect of the system clearly bears a close resemblance to that suggested by Kemp for the North Sea.

(4) Expenditures on 'exploration, development and construction' of a mine complex constitute capital investment. However, in determining the sum of investment expenditure which becomes the denominator of the ratio of profits to investment, each year's expenditure is 'grossed up' by an interest-rate factor until commercial production begins. The beginning of commercial production is defined as the date at which a mine reaches 60 per cent of its design output.

(5) Operating profit is defined as gross sales minus production costs and a number of other allowances, the most significant of which is the capital recovery allowance. For the first year of production the capital recovery allowance is precisely the figure for investment expenditure derived under para. (4). The important point, however, is that any unclaimed capital recovery allowance is grossed up by the interest-rate factor and is available for use in the following year. This process continues until the capital recovery allowance is exhausted, and effectively ensures that the full *present value* of capital investment is allowable against royalty payments and that no royalty is payable until this present value is recovered by the company.

(6) The interest-rate factor is the relevant year's average of the prime rate[63] plus 10 per cent of that rate (i.e. if the average rate were 10 per cent the interest rate factor would be 11 per cent).

The scheme is less complicated than it may appear at first. Suppose that investment of $K_0$ takes place in a single year and that production starts in the following year. Suppose also that net revenue in each year is $R_i$ where net revenue is gross revenue minus operating costs and all allowances except the capital recovery allowance. A hypothetical course of events is illustrated in Table 6.4 with $t$ representing the rate of tax applying to operating profit $R_4$.

TABLE 6.4

| Year | Capital recovery allowance | Net revenue | Operating profit | Tax |
|---|---|---|---|---|
| 0 | $K_0$ | – | – | – |
| 1 | $K_0(1 + r)$ | $> R_1$ | 0 | 0 |
| 2 | $K_0(1 + r)^2 - R_1(1 + r)$ | $> R_2$ | 0 | 0 |
| 3 | $K_0(1 + r)^3 - R_1(1 + r)^2 - R_2(1 + r)$ | $= R_3$ | 0 | 0 |
| 4 | 0 | $R_4$ | $R_4$ | $tR_4$ |

It is seen that no royalty is payable until year 4. Further, by dividing the entries in line 4 (year 3) by $(1 + r)^3$ and rearranging we obtain

$$K_0 = \frac{R_1}{(1 + r)} + \frac{R_2}{(1 + r)^2} + \frac{R_3}{(1 + r)^3}.$$

By failing to tax the net revenues $R_1$, $R_2$ and $R_3$ the company is assured of a rate of return $(r)$ on its investment. If $r$ is the company's rate of discount it also ensures that the project has a non-negative net present value. Only returns resulting in a positive net present value (or a rate of return greater than $r$) are taxed. The association between this scheme and the Garnaut and Clunies Ross suggestion is now apparent. In this case, however, progressivity is achieved not by recalculating Table 6.4 for different values of $r$ and applying additional rates of tax, but simply by taxing operating profit, when it occurs, at a progressive rate related to the ratio it bears to capital investment (in the hypothetical case above $K_0(1 + r)$).

The 'grossing up' of the capital recovery allowance year by year in the way described and the exemption from graduated royalty of operating profits below 15 per cent of capital expenditure clearly reflect the authorities' concern not to discourage marginal projects, while linking the tax rate to the ratio of profit to investment avoids the discrimination against large projects which a progressive tax regime might otherwise entail. Indeed if net revenues were expected to remain positive throughout the life of the mine, if the interest rate used were equal to the firm's rate of discount, and if the tax rate were simply proportional to operating profit the tax would amount to a proportional tax on net present value and hence would be allocatively neutral.

In the case of our example the net present value of the project can be represented by

$$V = \frac{R_4}{(1 + r)^4} + \frac{R_5}{(1 + r)^5} + \cdots + \frac{R_N}{(1 + r)^N},$$

where $N$ is the project life. Taxing the $R_i$ at rate $t$ will leave a post-tax net present value ($V_t$) of

$$V_t = (1 - t) \left[ \frac{R_4}{(1 + r)^4} + \frac{R_5}{(1 + r)^5} + \cdots + \frac{R_N}{(1 + r)^N} \right]$$

Clearly the capital expenditure and revenue flow which maximise $V$ will also maximise $V_t$ so that no distortion of choice is involved.

These conditions do not apply in practice, however, and some reallocation of resources would be expected to occur. In particular, with a progressive rate structure there will, in theory, be some incentive to adopt more capital per unit of output or to reduce the rate of exploitation for any given level of capital investment. This will not only delay the date at which tax becomes payable but will result in a lower average rate of tax applied to the net revenues since they will bear a smaller proportion to capital investment. A sufficiently progressive tax schedule might then induce 'wasteful' capital expenditure or a slower rate of output. The quantitative significance of this effect in the case of the graduated royalty is impossible to predict *a priori*, although the fact that the top rate of tax is 50 per cent suggests that the authorities wished to avoid it. On the other hand, a top rate of 50 per cent implies that a large proportion of the rent on particularly rich deposits of uranium will remain with the mining companies. The conflict between a large government 'take' (progressivity) and neutrality is therefore still in evidence.


## 6.7 Some Issues in North American Tax Policy

Tax policy towards natural resource industries in North America is a large and complex topic in its own right, and only a brief survey can be offered here.[64] It is hoped that some of the important issues can be identified and related to the theoretical discussion of Sections 6.2 and 6.3. Controversy centres around a number of important tax provisions – percentage depletion, 'expensing' and the treatment of capital gains.

As explained in Section 6.2 percentage depletion permits a company to claim as a tax allowance a certain percentage of the gross value of output. For integrated oil and gas producers (i.e. those with refineries or retail outlets), percentage depletion was abolished by the U.S. Tax Reduction Act of 1975. It remains intact, however, for other minerals such as uranium, where the rate is 22 per cent. It also remains, though at a reduced rate, for unintegrated small producers of oil or gas.[65]

The original intention of the Internal Revenue Act 1913, which introduced the Federal Income Tax, was that mineral producers should be permitted a deduction to allow for the depletion of their resources, and that the total deduction, as with other business assets, should equal the cost of acquiring them. During the First World War, however, a new principle began to gain acceptance, that the depletion allowance should be related to 'discovery value' – a value which for a successful venture could be greatly in excess of the exploration and development expenditures incurred. Percentage depletion, introduced in 1926, was intended to be a 'rule of thumb' for estimating the required allowance after it had become apparent that ascertaining discovery value was subject to much uncertainty and liable to give rise to interminable disputes.[66] The end-result was that total allowances available under percentage depletion greatly exceeded the sums which would be implied by a 'cost-depletion' regime. Mineral industries

were favoured still further by the ability to 'expense' development and exploration expenditure. In the case of oil and gas,
'intangible' costs of drilling may be expensed as soon as
production begins, as we have seen, so that a large proportion
of a firm's outlay on bringing a deposit into production is
allowed against tax immediately. This fact has no effect on
the percentage depletion deduction available, however, and
mineral firms are therefore effectively receiving two allowances for the same investment – the so-called 'double-dip'.
The tax benefits from these two provisions were estimated
as $2.6 billion in 1977.[67]

The provisions of the capital gains tax also tend to favour
the owners of energy resources. In the case of an owner of
industrial machinery it may happen on occasion that, in the
event of sale, a price will be realised considerably in excess of
that expected on the basis of its historic cost and the depreciation allowances permitted for tax purposes. Where such an
event occurs, the gain (up to the amount of depreciation
previously taken) is taxed as if it were ordinary income to the
firm, i.e. at 48 per cent. In the case of natural resources,
however, any gain in the value of a property on sale is taxed
at the lower capital gains tax rate of 30 per cent – irrespective
of the investment outlays 'expensed' or any percentage
depletion allowance taken. A further advantage stems from
the fact that abortive exploration expenditure may be
written off immediately against ordinary income. The Government therefore shares 48 per cent of the losses but only 30
per cent of the gains on exploration activity, a circumstance
likely to favour riskier undertakings as Page emphasises.[67]

Similar advantages of 'expensing' and percentage depletion
are accorded to the Canadian extractive industries. Percentage
depletion in this case however amounts to $33\frac{1}{3}$ per cent of
the taxable income rather than gross income and, as in the
United States, the privilege has been curtailed in recent
years.[69] In the light of these tax provisions debate in North
America has centred around their likely effects and objectives.
Unlike the specific instances of the petroleum revenue tax in
the United Kingdom and the uranium royalty in Canada,
which whether they succeed or not, are aimed at clear policy
objectives, the fiscal measures described above have tended

to evolve gradually as the result of changing pressures and circumstances so that their underlying rationale is obscure. The dominant issues have concerned the effects of the tax advantages on efficiency in resource allocation, and their impact on national security.

*Efficiency Implications*

All commentators are agreed on one point, percentage depletion and the other provisions, by lowering the effective rate of tax on the discovery and extraction of minerals relative to other activities, encourage capital to move into the favoured occupation. The precise ways in which this in turn effects crucial variables such as mineral prices, depletion rates, asset prices, exploration activity and so forth is still a matter of controversy. In Section 6.2 the impact of various tax provisions was investigated using a simple model of depletion and it is not intended to recover that ground here. We noted, however, that all tax advantages were likely to raise the value of resource deposits and hence rents or 'royalties', while capital gains privileges and percentage depletion might be expected to pull in opposite directions, the former raising present prices and retarding depletion, and the latter reducing present prices and advancing depletion.

During the 1960s opinion on the efficiency effects of the tax system divided broadly into two schools. On the one hand it was argued[70] that efficiency in resource allocation required *before tax* rates of return to be equal across sectors and that special privileges to the minerals industry induced capital to move from more to less socially productive purposes until *after tax* rates of return were equalised. On the other hand it was asserted that percentage depletion could be regarded as a way of correcting for the distortions introduced by the corporation tax itself.[71] In essence the argument was that the corporation tax represented a tax on capital which was fully shifted forwards to consumers. The more capital intensive the industry the greater the tax paid per dollar of output and the greater the percentage rise in product price necessary to fully pass forward the burden. The petroleum industry was just such a capital intensive industry, and the percentage

depletion allowance which could be regarded, as we have
seen, as a sales subsidy was required to restore petroleum
prices to their pre-corporation tax levels relative to the prices
of other goods. This ingenious view of percentage depletion
as being consistent with a 'neutral' tax system gave rise to
considerable comment – that the forward shifting assumption was too strong, that the observed capital intensity of
petroleum extraction was itself due to state regulations which
encouraged overdrilling, and that the tax benefits were more
likely to result in higher rents than lower prices.[72]

From the perspective of more recent developments in the
theory of efficient taxation, however, it may be that the focus
of debate was misplaced. Lump-sum taxes and rent taxes
apart, there are sure to be efficiency losses generated by the
raising of revenue. The question is, given the sums to be
raised, and given the constraints under which the Government
is operating, what tax structure will minimise these efficiency
losses? It transpires that where there are constraints on the
ability of the Government to tax all commodities (one
obvious example is the difficulty of taxing leisure) an optimal
tax structure will involve loading a relatively heavy burden on
those commodities for which there is a relatively inelastic
demand. We have in fact already met this result in Chapter 4
(p. 105) when discussing public utility pricing in the presence
of a financial target. Thus an efficient tax regime does not
necessarily imply that relative product prices should not alter,
and a relatively heavy tax on certain commodities may be
quite consistent with efficiency. Neither is it necessary for
capital to be taxed at the same rate in every sector. As
Stiglitz[73] argues, where commodity taxes cannot be varied
sufficiently between sectors a capital tax may act as a substitute. Where elasticities of substitution between capital and
labour are very low a tax on capital will approximate a tax on
output. Thus in this particular situation an efficient tax
regime, given the constraints, may involve taxing capital at
differential rates in different sectors, with higher rates applying to those sectors facing low demand elasticities and with
low levels of capital intensity. Where precisely this leaves the
petroleum industry is an unsettled question, although Stiglitz
suggests that there is little reason on grounds of efficiency to

tax the industry at lower rates than in manufacturing. In any event the justification for 'expensing' and percentage depletion as aids to economic efficiency is thus seen to depend on more than relative capital intensity.

A somewhat related question concerns the riskiness of minerals exploration and development. An important strand in the argument for a lower tax rate on mineral projects is that they face unusual risks. It is certainly true that for a given exploration project the 'risk' as measured by the standard derivation (say) of possible returns may be very great. But there are at least three reasons why economists are generally reluctant to accept the case that such risky undertakings require encouragement.

(*a*) As we saw in Section 6.3 a sufficiently diversified portfolio of risks, i.e. through risk-pooling or risk-spreading will reduce the overall risk of mineral company operations.

(*b*) If there are institutional constraints on the ability of markets to spread risks, constraints which cannot be overcome by improvements in insurance markets or markets in financial assets, there is no case on efficiency grounds for subsidising risk-taking. Risk which is borne by individuals is a cost of production. The fact that in an ideal world this cost might be lower through greater provision for risk-spreading is of no consequence for public policy if, in the nature of things, the ideal cannot be achieved.

(*c*) Where the tax code makes provision for loss-offsets (i.e. taxes are paid on gains but may be reclaimed on losses) the corporation tax is likely to stimulate risk taking in any case.[74] This is all the more probable, as we have already noted, if the Government as a 'silent partner' shares a greater proportion of the losses compared with the gains.

## National Security

In Section 6.3 the desire for security was identified as an important influence on the allocation of resources in the field of energy. Given the difficulties of explaining the special tax provisions by reference to efficiency considerations the other possibility is that they are aimed at providing the United States with sufficient 'reserves' to prevent any foreign supplier becoming a threat to national independence. The issue, which has been extensively debated in the United States, may be divided into two. First it is necessary to know whether the tax concessions concerned actually increase proven petroleum reserves, and second, if so, whether the method is cost-effective relative to alternatives such as quotas, tariffs and strategic stockpiles. A detailed appraisal of these different instruments would take us beyond the scope of this chapter.[75] However, empirical work undertaken in the United States[76] indicates that the percentage depletion allowance and the expensing of intangible drilling costs do increase the investment of the petroleum industry in proved reserves, but that these tax measures appear to be an expensive method of achieving the objective.

## The Tax Reduction Act 1975

Increasing scepticism about the relevance of the percentage depletion allowance for achieving any desired objective, and indeed the general view that it led to inefficient resource allocation played an important part in its partial removal in 1975. Concern about the use of special tax preferences to shelter large amounts of income had already led in 1969 to the introduction of the 'minimum tax'. By 1976 this tax was levied at a rate of 15 per cent on preference items above an exemption level of $10,000 or regular income tax paid, whichever was greater. Thus percentage depletion in excess of cost depletion; 18/48ths of long-term capital gains (a 30 per cent tax rate on capital gains is equivalent to a full 48 per cent on 30/48ths of the gains thus implying that 18/48ths are untaxed); and the excess of intangible drilling costs expensed over the allowance permitted if they were depreciated under normal procedures, are all preference items on which the 15 per cent tax is levied. The removal of the percentage depletion allowance for integrated oil and gas producers can thus be seen as part of a general 'tightening up' of tax allowances and not merely as a move dictated solely by considerations of energy policy.

The effects of removing the percentage depletion allowance should be equivalent to introducing an *ad valorem* severance tax analysed in Section 6.2. Present prices will rise, royalties will fall and depletion will be delayed. Alternatively, the analysis can be conducted using the more traditional approach of Figure 6.1. Given that the OPEC price defines an upper limit for domestic supplies, abolishing the allowance shifts the supply curve to the left and, as we saw in Section 6.3, implies a rise in imports. In itself, therefore, removing percentage depletion would seem to be inimical to the achievement of a national security objective defined in terms of a limitation of imports. A tariff or quota which raised import prices would restore the situation and the policy package would then represent the substitution of a price-incentive for tax-incentives to the U.S. petroleum industry. It is this analysis which led McDonald[77] to remark that 'most Americans will gain as taxpayers what they lose as consumers of higher priced oil and gas'.

## 6.8   Conclusion

The use of tax and subsidy instruments in the field of energy policy is a very extensive subject, and inevitably we have had to focus on a few major areas. It is evident from our analysis, however, that the achievement of even a fairly straightforward objective such as the taxation of excess profits or rents from minerals exploitation is fraught with considerable difficulty. When combined with additional objectives relating to the rate of development and exploration, the rate of depletion, national security, distributional equity and so forth, the problem is complicated still further. This complexity may go some way to explain the increasing use of regulatory rather than purely fiscal instruments. In spite of the possibility of introducing measures such as those discussed in Sections 6.2 and 6.3, the U.K. Government has on the whole opted for direct control. Thus we noted in Section 6.5 the preference for ministerial discretion in the allocation of licences compared with competitive bidding, the use of 'participation agreements' rather than exclusive reliance on the petroleum revenue tax, and the taking of direct power to control depletion rates[77] instead of introducing an appropriately designed severance tax.

Conditions in North America provide a contrast although even here price controls play an important role as was noted in Section 6.4 (see also Chapter 9, Section 9.4). In general, however, much greater reliance is placed on lease bonuses to appropriate rent from the development of resources on Federal land or on the continental shelf, and recent changes in tax policy described in Section 6.7 have been specifically aimed at reducing the rate at which indigenous sources of supply are depleted.

# 7. UNCERTAINTY AND ENERGY POLICY

## 7.1 Introduction

At many points in earlier chapters we have had occasion to draw the reader's attention to the problem of uncertainty. Uncertainty concerning stocks of fossil fuels, future price trends, technical changes and the appearance of a 'backstop technology' were referred to in Chapters 2 and 3. In Chapter 4 uncertainty about future demand for energy was observed to create problems for pricing and investment policy. Uncertainty about the hazards associated with various environmental pollutants is clearly important in the context of the analysis in Chapter 5, while uncertainty about future price and cost trends and the probability of discovering deposits of fossil fuel was observed to play a large part in determining policy towards taxation in Chapter 6.

In this chapter a more systematic attempt is made at drawing some of these issues together. The objective is to outline the various theoretical models of uncertainty developed by economists, and to comment on the extent to which they can be used to assist decision-making. Inevitably much of the chapter will appear to be of more theoretical than practical interest. However, the problem of coping with uncertainty is one of the most difficult that economists face. As William Vickrey has written: 'while choice among alternatives is an everyday matter, it is the particular province of political economy to choose among methods of choice'. Only through the development of theoretical models can the full implications of various choice criteria be appreciated and their various strengths and weaknesses be identified.

It is common to distinguish situations of risk from situations of uncertainty. In the former case the decision-maker is assumed to have some knowledge about the probabilities attached to the possible outcomes, whereas in the latter case no such knowledge is presumed. Although widely accepted it is not usual to stick slavishly to these definitions and in the following pages the term 'uncertainty' will be used to describe both situations.

The rest of this chapter is divided into eight further sections. Section 7.2 introduces the state-preference approach to uncertainty and derives some basic results. In the following section some space is devoted to further illustrating the concepts involved by the use of a simple numerical example. In Section 7.4 the applicability of the Arrow–Lind theorem to the analysis of energy projects is discussed, and Section 7.5 provides an explicit treatment of 'irreversibilities' which are of considerable importance in some areas of energy policy. A short general critique of the state-preference approach follows in Section 7.6 and Section 7.7 introduces alternative approaches to decision-making which do not rely on knowledge concerning probabilities. Section 7.8 attempts to highlight the philosophical political and ethical questions raised by the earlier theorising. In particular, consideration is given to the derivation of subjective probability estimates when relative frequency information is unavailable, and the use of 'experts' in providing these estimates, setting standards and collecting information. Some concluding comments make up Section 7.9.

## 7.2 Markets and Uncertainty

This section attempts to provide a brief introduction to some of the concepts and methods which have been developed in the economics literature to cope with the problem of uncertainty.[1] It is assumed at the outset that some objective and agreed upon probabilities can be assigned to the occurrence of various events, and the immediate aim is to discover how a perfectly operating market system might be expected to respond to these circumstances. Even at this early stage it might be questioned how relevant such a model is likely to be to the understanding of resource allocation in the field of energy. However, it is introduced here not as a description of the way the world operates, but as an aid to the understanding of certain important results which will be developed in later sections.

A prerequisite of any theory of individual behaviour is a statement of the objectives which it is assumed the individual is pursuing. In the traditional case of consumption behaviour under certainty the individual is supposed to maximise a utility function defined over bundles of consumption goods. The generalisation of this case to inter-temporal choice, which was explored in Chapter 3, entailed modifying the utility function to take account of the existence of several time periods. A further generalisation to the case of uncertainty is more difficult, but formally at least we might proceed by modifying the utility function once more, this time allowing for the existence of several possible 'states of the world'. Suppose, for example, an individual is planning his consumption decision but is unsure which of two mutually exclusive states of the world exists. His utility function could then take the general form

$$V^* = V^* (C_{1a}, C_{1b}, \Pi_{1a}, \Pi_{1b}), \qquad (7.1)$$

where $C_{1a}$ = consumption in period 1 in the event of state of the world $a$ occurring and $\Pi_{1b}$ = probability of state of the world $b$ occurring in period 1. Note that either state $a$ or state $b$ must occur, but not both; for example, the individual may win money on a public lottery or he may not. It follows that the arguments in the utility function do not refer to ex post consumption levels. If state of the world $a$ occurs (the individual wins a bet) then he will actually consume $C_{1a}$. Utility, however, is dependent here not on the actual outcome, since this cannot be known, but on the distribution of all possible outcomes.

Clearly this formulation is somewhat vague, and it is desirable to know more details about the form of the utility function. One suggestion might be that individuals are interested in maximising the expected value ($V$) of the outcomes where

$$V = \Pi_{1a} C_{1a} + \Pi_{1b} C_{1b}. \qquad (7.2)$$

This simple proposition can be shown, however, to have certain 'unreasonable' implications. In particular it implies that an individual will always be willing to accept a favourable bet, i.e. a bet with a positive expected value, and will be indifferent between taking and not taking actuarially 'fair' bets. Such a person is described as being 'risk-neutral'.

As an example, consider a game involving the toss of a fair coin. Our individual is invited to participate and will receive £6 if a 'head' occurs but must pay £2 in the event of a 'tail'. The expected value of the bet is therefore $(\frac{1}{2} \times 6) - (\frac{1}{2} \times 2) = £2$ and the individual will hasten to play the game. Indeed he would be willing to pay up to £2 for the privilege of participating, since at any sum less than this the game will still add to the expected value of his consumption claims. Where small sums of money are involved, or where such a game can be repeated, very large numbers of times such 'risk-neutrality' may not appear startling. A casino survives by offering just such 'fair' or slightly 'unfair' bets for relatively small sums. Severe problems arise, however, when large variations in outcomes are possible and the game is to be played only once. The classic example of this situation is that of the 'St Petersburg Paradox', analysed by Daniel Bernoulli, the eighteenth-century mathematician.

Consider once more the tossing of a fair coin. Suppose that the individual's winnings depend on how many times the coin must be tossed before a 'head' appears. Further let the winnings be £2n, where $n$ is the number of trials required. Given that the chance of obtaining the first 'head' on the first toss is $\frac{1}{2}$, on the second toss $\frac{1}{4}$, on the third $\frac{1}{8}$, etc., the expected winnings from this game will be given by the following expression:

$$(\tfrac{1}{2} \times 2) + (\tfrac{1}{4} \times 2^2) + (\tfrac{1}{8} \times 2^3) + \cdots + \left(\frac{1}{2^n} \times 2^n\right) + \cdots,$$

Which sums to infinity. It seems inconceivable that anyone would be willing to pay even £1000 for a single game much less an infinite amount. The random variable upon which 'winnings' depend is in fact geometrically distributed with the expected number of trials required to obtain a head equal to two!

The implication of an unwillingness to commit substantial funds in return for playing this game is obviously that the infinitesimally small probability of obtaining astronomically large gains does not compensate for the near certainty of substantial losses. Individuals, it appears, are not concerned with the pay-off which various states of the world imply simply in terms of money or consumption claims. They are interested in the satisfaction or utility which that consumption will confer if experienced. Assuming that the marginal utility of consumption declines as consumption rises, it then follows that a given loss will be felt more keenly than a gain of the same amount, and that even an actuarily 'fair' bet in terms of money will result in a loss when evaluated in terms of 'expected utility'. Thus, if we assume that individual decision-makers wish to maximise 'expected utility' and that they have utility functions exhibiting diminishing marginal utility, they will reject fair bets and hence will be 'risk-averse'. The new formulation of the individual's objective function is therefore

$$E(V) = \Pi_{1a} V(C_{1a}) + \Pi_{1b} V(C_{1b}) \qquad (7.3)$$

with $\quad V'(C_{1a}), V'(C_{1b}) > 0$

$$V''(C_{1a}), V''(C_{1b}) < 0.$$

It might be objected that this objective function is arbitrary and that there is no more reason to suppose that consumers maximise expected utility than expected money returns. They might be concerned with the variability of utility outcomes and not simply the expectation. It can be shown however, that providing the consumer's preferences accord with certain axioms of choice – the von Neumann and Morgenstern axioms – a utility function can be constructed which permits the prediction of choice under uncertainty on the basis of expected utility.[2]

Figure 7.1 illustrates the case of a risk-averse consumer with utility function $V(C_1)$. Suppose that this individual is able to consume $\bar{C}$ with certainty, but also has the opportunity of taking a fair bet with a chance $\Pi_{1a}$ of increasing his consumption level to $C_{1a}$ and a chance $\Pi_{1b} = 1 - \Pi_{1a}$ of finishing up with $C_{1b}$. Since the bet is 'fair' the expected

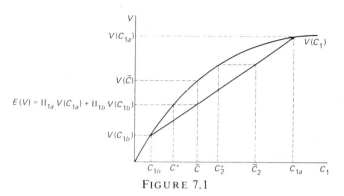

FIGURE 7.1

consumption of the individual remains at $\bar{C}$. This, however, is a mathematical statement and the consumer knows that if he takes the bet $\bar{C}$ cannot actually occur. This uncertainty over what the final outcome in fact will be represents a cost to a risk-averse individual, and it is seen from the figure that the expected utility of the bet $\Pi_{1a} V(C_{1a}) + \Pi_{1b} V(C_{1b})$ is less than the utility derivable from the certain option $V(\bar{C})$. Providing that the utility function is concave, as in Figure 7.1, it is evident that the consumer will always reject a fair bet.

A corollary of this observation is that a risk-averse individual will always be willing to take 'fair' insurance. Again with reference to Figure 7.1, if the consumer starts out facing the uncertain prospect he will be willing to pay the amount $\bar{C}C_{1a}$ to an insurance company in return for the amount $C_{1b}C_{1a}$ in the event of the unfavourable state $b$ occurring. In this way he will be sure of a consumption level $\bar{C}$ whichever state of the world prevails and his utility will increase from $E(V)$ to $V(\bar{C})$. For its part the insurance company would be taking a fair bet since it would stand to gain $\bar{C}C_{1a}$ with probability $\Pi_{1a}$ and lose $\bar{C}C_{1b}$ with probability $\Pi_{1b}$. The implied risk-neutrality of the insurance company can then be justified by pointing out that it pools many similar but independent risks; just as an individual may exhibit risk-neutral behaviour when a game of chance can be repeated many times.

From Figure 7.1 it is also possible to demonstrate the concept of 'certainty equivalence'. Suppose the individual faces the uncertain prospect of $C_{1a}$ or $C_{1b}$ with expected outcome $\bar{C}$ and expected utility $E(V)$. There will exist some amount of consumption which, if offered to the consumer with certainty, will confer the same utility as the uncertain gamble. This certain sum is termed the 'certainty equivalent' of the gamble and in the figure is represented by $C^*$. The distance $C^*\bar{C}$ – that is, the difference between the expected value of the gamble and its 'certainty equivalent' – can be regarded as the 'cost of risk-bearing', a concept which turns up in various disguises in the literature on uncertainty. Thus a consumer will be willing to purchase insurance even at slightly 'unfair' rates as long as the outcome is to ensure the enjoyment of a sum greater than the certainty equivalent of the gamble (i.e. remaining uninsured).

It is worth noting that risk-aversion implies a disinclination to take fair bets; it does not imply that a risk-averse person will never bet. Suppose, for example, that a consumer has a stock of consumption claims $\bar{C}$ but is offered a gamble still involving $C_{1a}$ and $C_{1b}$ but with the odds sufficiently improved to result in an expected outcome $\bar{C}_2$. From Figure 7.1 it is seen that the expected utility of the gamble exceeds $V(\bar{C})$ and the consumer, wishing to maximise expected utility, would take the risk. We could express the same thing in alternative terminology by saying that the certainty equivalent of the gamble, $C_2^*$ in the figure, exceeds $\bar{C}$.

## State Preference Theory

The paragraphs above implicitly represent an exercise in state-preference theory and it will prove useful to recast the analysis explicitly in this form. This will have the advantage of providing greater generality as well as illustrating the close association with more conventional consumer theory. So far we have looked closely at the problem of specifying an objective function for the individual under conditions of uncertainty but have yet to explicitly consider the constraints. From equations (7.1), (7.2) and (7.3) it is seen that the individual's utility depends upon the probability of various states of the world occurring and the consumption that will be possible in those states. To carry the analysis further it is assumed that it is possible to purchase and sell claims to units of consumption or income in all future time periods and contingent upon the occurrence of given states of the world. These claims, which are tradeable on present markets, are called 'state-contingent claims' and will have market determined prices $P_{1a}, P_{2b}$, etc. Thus $P_{2b}$ would be the price today of a claim to £1 in period 2 contingent upon the occurrence of state of the world $b$. If state of the world $a$ occurs the claim will, *ex post*, turn out to be worthless.

## The Consumption Decision

Ignoring for the present the possibility of incorporating differing time periods and concentrating on the simple two-state example introduced earlier, the consumer's choice problem can be set up formally in the familiar way

$$\text{Max } \Pi_{1a} V(C_{1a}) + \Pi_{1b} V(C_{1b})$$

$$\text{subject to } W = P_{1a} C_{1a} + P_{1b} C_{1b}$$

where $W = P_{1a} Y_{1a} + P_{1b} Y_{1b}$ and e.g. $Y_{1b}$ = income in the event of the occurrence of state $b$. Thus $W$ is the equivalent of the income constraint in traditional consumer theory.[3] The Lagrangean function is simply

$$L = \Pi_{1a} V(C_{1a}) + \Pi_{1b} V(C_{1b})$$
$$+ \lambda(W - P_{1a} C_{1a} - P_{1b} C_{1b})$$

and the first-order conditions are therefore

$$\Pi_{1a} V'(C_{1a}) - \lambda P_{1a} = 0 \tag{i}$$

$$\Pi_{1b} V'(C_{1b}) - \lambda P_{1b} = 0 \tag{ii}$$

$$W - P_{1a} C_{1a} - P_{1b} C_{1b} = 0. \tag{iii}$$

From the first two of these conditions we then obtain

$$\frac{\Pi_{1a} V'(C_{1a})}{\Pi_{1b} V'(C_{1b})} = \frac{P_{1a}}{P_{1b}}. \tag{7.4}$$

Figure 7.2 illustrates this position of equilibrium with the consumer holding $\bar{C}_{1a}$, claims to consumption contingent on state of the world $a$ and $\bar{C}_{1b}$ claims contingent upon $b$. It can be shown that in the case of a function such as expression (7.3) with $V$ concave, indifference curves can be derived

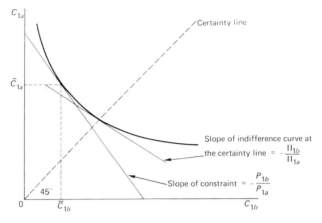

FIGURE 7.2

which are convex as in Figure 7.2 and the first-order conditions are sufficient for a maximum.

An important result follows from expression (7.4). If the function $V$ does not depend on the state of the world that is realised, as has so far implicitly been assumed, and if the relative prices of state contingent claims are equal to the ratio of state probabilities, the consumer will opt for certainty. Thus from expression (7.4) if

$$\frac{P_{1a}}{P_{1b}} = \frac{\Pi_{1a}}{\Pi_{1b}} \ , \ V'(C_{1a}) \text{ will equal } V'(C_{1b})$$

and hence $\bar{C}_{1a} = \bar{C}_{1b}$. The consumer will wish to ensure the *same* level of consumption whichever state of the world pertains. This result is, of course, equivalent to our earlier observation that risk-averse individuals would reject fair bets and always accept actuarially fair insurance.

A solution implying *unequal* consumption between the two states may occur if utility is itself contingent on the state of the world. An extreme example of this possibility would occur if the states were 'alive' and 'dead' respectively. A person with many dependants might still value fairly highly claims to consumption occurring in the event of his death, but someone with fewer family responsibilities would be expected to value such claims at very little. Expected utility should therefore more generally be written as

$$E(V) = \Pi_{1a} V_a(C_{1a}) + \Pi_{1b} V_b(C_{1b}).$$

The case illustrated in Figure 7.2 does not, however, depend upon state-contingent utility functions. Willingness to opt for a 'risky' solution is here the result of the ratio of claims prices diverging from the ratio of state probabilities. The latter ratio is given by the slope of the indifference curve at its point of intersection with the certainty line,[4] whereas the former ratio is given by the slope of the constraint. Thus, as drawn, we have

$$\frac{\Pi_{1b}}{\Pi_{1a}} < \frac{P_{1b}}{P_{1a}} \quad \text{or} \quad \Pi_{1b} \cdot \frac{1}{P_{1b}} < \Pi_{1a} \cdot \frac{1}{P_{1a}}.$$

Recollect that $P_{1b}$ is the price of a unit of consumption contingent upon state $b$ occurring. It follows that $1/P_{1b}$ will represent the number of contingent consumption units that can be purchased for £1. Hence $\Pi_{1b} \cdot 1/P_{1b}$ equals the expected number of consumption units purchased for £1 if state $b$ contingent claims are held. Similarly, $\Pi_{1a} \cdot 1/P_{1a}$ will be the expected number of consumption units purchased for £1 if state $a$ contingent claims are held. It is evident that from the point of view of *expected* returns state $a$ contingent claims are a better bet and this is reflected in a larger holding of such claims. Note that were the individual consumer risk-neutral and hence interested only in expected returns he would opt to hold state $a$ claims exclusively. Indifference curves in such a case would be straight lines and not convex curves, and a 'corner solution' would result. In the risk-averse case the consumer can be induced to move to less certain positions involving more $C_{1a}$ and less $C_{1b}$ by increasing $\Pi_{1a}$ and hence altering the shape of the indifference curve or by reducing $P_{1a}$ and hence steepening the slope of the constraint. This result accords with our earlier deduction from Figure 7.1 that the consumer could be induced to move from a certain position $\bar{C}$ to take a gamble, providing that the odds were sufficiently favourable.

### The Investment Decision

In Chapter 3 (p. 32) it was noted that when making inter-temporal consumption and production decisions the decision-maker will attempt to maximise the present value of his wealth, evaluated at the prevailing riskless rate of interest. In perfect capital markets with no uncertainty, therefore, the investment criterion is to accept a project if the net present value is positive, i.e.

$$W = -K_0 + \frac{R_1}{1+r} + \frac{R_2}{(1+r)^2} + \cdots + \frac{R_n}{(1+r)^n} > 0,$$
(7.5)

where   $W$ = net present value,

$Ri$ = net returns in period $i$,

$r$  = interest rate,

$K_0$ = initial investment outlay in period zero.

The same criterion can be expressed in rather different notation as follows:

$$W = -P_0 K_0 + P_1 R_1 + P_2 R_2 + \cdots + P_n R_n > 0, \quad (7.6)$$

where $Pi$ = present price of a claim to £1 in period $i$.

Clearly if an asset promises to pay the bearer £1 in the next period its price in a perfect market will be $1/(1+r)$. Similarly, if the payment date is not until two periods hence, its price will be $1/(1+r)^2$. Thus expressions (7.5) and (7.6) are entirely equivalent in these circumstances. Expression (7.6) is easier to generalise to the case of uncertainty, however, for in this case a market exists not simply in claims to be redeemed in specified time periods but also in claims redeemable only in specified 'states of the world'. Suppose, for example, a project involves investment now, i.e. period zero, and produces returns next period, i.e. period 1. Suppose also that these returns are dependent upon which of two mutually exclusive states of the world prevail when period 1 arrives. The investment criterion could in these circumstances be written

$$W^* = -P_0 K_0 + P_{1a} R_{1a} + P_{1b} R_{1b} > 0, \qquad (7.7)$$

where    $P_{1a}$ = price of claim to £1 in period 1 contingent upon state $a$,

$R_{1b}$ = net returns in period 1 contingent upon state $b$, etc.

If expression (7.7) is positive the individual investor will adopt the project. Since the project produces returns of $R_{1a}$ if state $a$ occurs, claims to these returns could be sold in the market for $P_{1a} R_{1a}$. Similarly, since returns of $R_{1b}$ will result if state $b$ occurs, claims to these can be sold now for $P_{1b} R_{1b}$. If the sum total of these receipts exceeds $P_0 K_0$, the required outlay, it is clearly worth while to proceed with the project. $W^*$ is termed the 'present certainty-equivalent value' of the project.[5]

It may appear strange that in expression (7.7) no explicit mention is made of discount rates or of attitudes to risk. These factors are taken into account by the fact that they determine the prices of the time and state-contingent claims. It can be seen, for example, that since either state $a$ or state $b$ must occur, we can only be certain of having £1 in period 1 if we possess two claims, one contingent upon state $a$ and the other on state $b$. Thus the present price of a certain pound in period one must be $P_{1a} + P_{1b}$. But we already know that this price equals $1/(1 + r)$, where $r$ is the riskless rate of interest. Hence we deduce that

$$P_{1a} + P_{1b} = \frac{1}{1 + r}. \qquad (7.8)$$

Further it would appear reasonable, if rather trivial, to suppose that $P_0 = 1$.

With the theoretical apparatus now available it is possible to make some preliminary comments about the appropriate rate of discount to apply under conditions of uncertainty to the evaluation of a given investment project. It has been seen that in the 'textbook' conditions assumed thus far, with perfect markets operating in state-contingent claims, the appropriate investment criterion is given by expression (7.7). We now define the 'risky' rate of discount $(p)$ as that rate which, when used to discount the *expected returns* from an investment project, produces the correct present certainty-equivalent value. Suppose, for example, that state $a$ will occur with probability $\Pi_{1a}$ and state $b$ with probability $\Pi_{1b} = 1 - \Pi_{1a}$. In this case the expected return in period one from the investment project will be given by

$$E(R_1) = \Pi_{1a}R_{1a} + (1 - \Pi_{1a})R_{1b}.$$

Discounting expected future returns at rate p gives a present value $(V)$ as follows

$$V = -K_0 + \frac{\Pi_{1a}R_{1a} + (1 - \Pi_{1a})R_{1b}}{1 + p}$$

The discount rate $p$ is to be chosen so that $V = W^*$, hence

$$-K_0 + \frac{\Pi_{1a}R_{1a} + (1 - \Pi_{1a})R_{1b}}{1 + p} = -K_0 + P_{1a}R_{1a} + P_{1b}R_{1b},$$

or $$1 + p = \frac{\Pi_{1a}R_{1a} + (1 - \Pi_{1a})R_{1b}}{P_{1a}R_{1a} + P_{1b}R_{1b}} \qquad (7.9)$$

Thus in this simple two-state two-period example the investment criterion (7.7) is equivalent to discounting expected returns at a rate $p$ given by equation (7.9). It is evident, however, that general statements about the magnitude of $p$ are difficult since each project will have its own particular state-contingent revenue flows $R_{1a}, R_{1b}$ and hence its own particular 'risky' rate of discount $p$. A few special features of equation (7.9) nevertheless deserve attention.

(i) Clearly if $R_{1a} = R_{1b}$, that is, if returns to the project are independent of states of nature $a$ and $b$, expression (7.9) becomes simply

$$1 + p = \frac{1}{P_{1a} + P_{1b}} = 1 + r \text{ (from (7.8))}.$$

Hence in these circumstances $p = r$. Not surprisingly perhaps, if the project is riskless it will be evaluated at the riskless rate of interest.

(ii) If two projects are being compared and each project has the same relative pay-off in the various states, the same rate of discount will be applied. Thus if $R_{1a}^* = kR_{1a}$, and $R_{1b}^* = kR_{1b}$, where $R^*$ and $R$ represent state-contingent returns in two different projects and $k$ is some constant, equation (7.9) will give the same 'risky' rate of interest for both projects. The two projects would be said to be in the same 'risk class'.[6]

(iii) Suppose that the ratio of the prices of state-contingent claims is equal to the ratio of state probabilities, i.e. let

$$\frac{P_{1a}}{P_{1b}} = \frac{\Pi_{1a}}{\Pi_{1b}}.$$

It follows that $\Pi_{1a}/P_{1a} = \Pi_{1b}/P_{1b} = k^*$ where $k^*$ is a constant, and hence $\Pi_{1a} = k^*P_{1a}$, $\Pi_{1b} = k^*P_{1b}$. Substituting for $\Pi_{1a}$ and $\Pi_{1b}$ in equation (7.9) we obtain

$$1 + p = \frac{k^*P_{1a}R_{1a} + k^*P_{1b}R_{1b}}{P_{1a}R_{1a} + P_{1b}R_{1b}} = k^*.$$

But          $\Pi_{1a} + \Pi_{1b} = k^*P_{1a} + k^*P_{1b} = 1.$

Hence     $k^* = \dfrac{1}{P_{1a} + P_{1b}} = 1 + r.$ Thus we find that

$$1 + p = k^* = 1 + r$$

and the 'risky' rate of discount $p$ equals the 'riskless' rate $r$. Where claims prices are proportional to state probabilities, investment criterion (7.7) therefore implies the use of a riskless rate of interest to discount expected returns. The reason behind this result can better be understood by considering once more expression (7.4). If $\Pi_{1a}/\Pi_{1b} = P_{1a}/P_{1b}$ the first-order conditions for the consumer imply that the marginal utility of consumption is the same whichever state of the world occurs. The expected utility conferred by the returns of the investment project can therefore approximately be written

$$E(V) = \Pi_{1a}V'(R) R_{1a} + \Pi_{1b} V'(R)R_{1b},$$
$$= V'(R) [\Pi_{1a}R_{1a} + \Pi_{1b}R_{1b}],$$

where $V'(R)$ = marginal utility of returns. Clearly the maximisation of expected utility in these circumstances implies the maximisation of expected returns and, as has already been seen, this in turn implies risk-neutral behaviour. The above expression is approximate in the sense that the investment project must be assumed to be sufficiently small not to alter the marginal utility of the returns.

(iv) Finally we inquire what are the consequences for $p$, the risky rate of discount, if claims prices and state probabilities are *not* proportional. Suppose that $\Pi_{1a}/P_{1a} < \Pi_{1b}/P_{1b}$. This implies that the price of state $b$ claims is rather lower and the price of state $a$ claims rather higher than would be expected merely on the basis of their respective probabilities. One explanation could be that utility functions differ between states, in this case resulting in a preference for consumption in state $a$. Another possibility is that the two states imply greatly differing consumption prospects. Thus state $a$ might be 'famine' and state $b$ a 'good harvest'. Whatever the reason, intuition and expression (7.4) confirm that the marginal utility of returns in state $a$ exceeds the marginal utility of returns in state $b$. If investment projects are being undertaken, people would prefer the returns to be loaded in favour of state $a$ rather than state $b$, as witnessed by the relatively high price they are willing to pay for state $a$ contingent claims. It transpires therefore that any investment project which accords with these preferences, that is for which $R_{1a} > R_{1b}$ will be evaluated at a 'risky' rate of discount *lower* than the riskless rate, while any project which produces returns predominantly in the lower-valued state will be evaluated at a rate of discount *greater* than the riskless rate.

This completes our excursion into state-preference theory. Individuals are considered to be risk-averse and to maximise expected utility rather than expected money returns. Risk to the individual can be reduced through the buying and selling of time and state-contingent claims. A perfectly operating market in these claims may in special cases avoid private risk altogether (marginal utility of consumption being constant whatever 'state of the world' occurs) and the present certainty-equivalent value criterion for investment projects would then imply the use of a riskless rate of interest to discount expected returns whatever the distribution of these may be between the various states. Where some 'social risk' cannot be avoided, as in the case of drought or famine, a risky rate of discount is implied which may be larger or smaller than the time-preference rate or riskless rate, according to the state-contingent distribution of returns from the project concerned. Where the project itself is riskless or, as will be seen later, where the returns are statistically independent of the returns from existing investments, the riskless rate of interest is still implied.

## 7.3  Exploration for Oil – A Numerical Example

Some of the issues raised in Section 7.2 may be clarified by the use of a simple numerical example. This example is not meant to be 'realistic' but does help to illustrate some important principles. Suppose that an oil company is about to embark on a project of exploratory drilling in a particular area offshore. Suppose also that the probability of striking oil in commercial quantities and the probability of all holes being dry are both one-half. In the event of a successful outcome the present value of returns will be £44 million when evaluated at a riskless discount rate. Where no oil is discovered a loss of £19 million will be experienced. The decision-makers at the company, whether private owners, top executives or majority shareholders, are assumed to be risk-averse and to have a utility of wealth function $V = \sqrt{W}$. The existing value of the company is £100 million which for simplicity is assumed to be certain. We ignore the difficult and important questions of how a firm's utility function might be derived from the individual preferences of managers or owners, or indeed why a firm *should* be risk-averse if shareholders hold diversified portfolios of claims, as discussed in Section 7.2. Risk-averse behaviour, however, would appear to be quite consistent with modern 'managerial' theories of the firm.[7]

Figure 7.3 reproduces the information given above. If the project is undertaken, expected wealth will be £112.5 million,

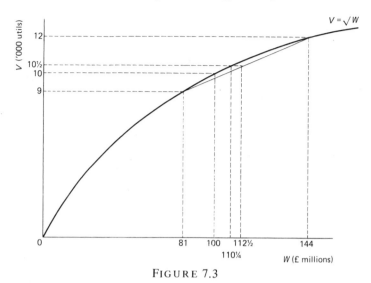

FIGURE 7.3

i.e. the project is expected to increase the value of the company by £12.5 million. $E(V) = 10,500$ and the certainty equivalent of the risky project is £110.25 million. Since this exceeds £100 million we deduce that the company will go ahead with the exploration activity. The cost of risk as defined in Section 7.2 (i.e. expected wealth minus its certainty equivalent) will be £(112.5 − 110.25) million = £2.25 million.

Now consider the same company evaluating the very same investment opportunity, but this time in the knowledge that it must purchase an exploration licence from the Government. In Chapter 6 it was noted (p. 168) that under conditions of risk the sale of licences could discourage exploration and we are now in a better position for seeing why this is so. Suppose the Government, aware that the expected return is £12.5 million, charges £12 million for the licence. This will still leave the project with a positive net present value when evaluated at the riskless rate of discount. However, the certainty equivalent of this new gamble, the outcomes of which are now £(100 − 31) million and £(100 + 32) million will be only £97.9 million and the firm will reject it. Being risk-averse the firm does not undertake all projects with a positive present value at the riskless rate of interest, it assesses the risks involved and makes some adjustment for them in its calculations. This adjustment may take the form of a higher discount rate than the riskless one for discounting expected future net benefits, a downward revision of expected net present value figures, the adoption of a minimum 'pay-back period' and so forth, but whatever the method the effect is to render unacceptable the exploration project. Some simple, if tedious, calculations will soon confirm that the maximum the firm would be willing to pay for a licence is approximately £10 million. Again with reference to Chapter 6 it is of interest to note that a system of royalty bidding, requiring the firm to pay only in the event of oil being discovered, would produce different results. In this case the firm would be prepared to offer £23 million, the expected value to the Exchequer therefore being £11.5 million.[8] Assuming risk-neutrality on the part of the Exchequer, a position for which, as will be seen, there is some theoretical justification, this would appear to be a preferable system. The result is achieved, of course, because the Government shoulders some of the risks of the project. In the former case the firm requires £2.5 million as compensation for risk-taking, in the latter case this figure has dropped to £1 million.

It was further remarked in Chapter 6 (p. 167) that one important criticism of the competitive bidding solution to the allocation of exploration licences was that under conditions of risk it relied upon the existence of efficient capital markets. Suppose a perfect market in state-contingent claims existed. The firm in our example could in this case insure against an unfavourable outcome. It could do so by selling claims to £31.5 million contingent upon oil being discovered and, with the proceeds, purchasing claims to £31.5 million contingent upon no oil being discovered. If claims prices were proportional to state probabilities both sets of claims would have the same price (since the probability of both events is 0.5) and the firm could thereby achieve a *certain* return of £12.5 million. This sum would then represent the maximum the firm would be willing to pay for an exploration licence, and the firm would have managed to avoid risk completely.

Finally, this numerical example provides an opportunity to illustrate the effects of risk-spreading. The possibility of risk-pooling and risk-spreading was mentioned in Chapter 6 but no detailed attention was paid to the distinction between them. Thus risk-pooling occurs when a given investor undertakes many independent risks, or risks which are dependent in such a way as to reduce the overall risk associated with the total pool. Risk-pooling is the principle underlying the popular saying that 'you should not have all your eggs in one basket'. Risk-spreading, on the other hand occurs, when the risk from a particular project is shared by several individuals.

In the particular case that is being considered in this section the private cost of risk has been calculated already as £2.25 million in the absence of insurance possibilities and assuming no exploration licence is required. Now suppose that another identical firm with the same utility of wealth function offers to share equally in the costs and revenues of the project. Each firm now faces a gamble, with outcomes of £90.5 million and £122 million. The expected outcome will be £106.25 million for both firms; that is, they both expect to add £6.25 million to the present value of their profits. This is unsurprising. The project as a whole still has an expected return of £12.5 million and the two firms now share this expectation between them. A more interesting conclusion emerges, however, when we look at the cost of risk-bearing associated with the project. For each firm the certainty equivalent of this new gamble is £105.66 million and hence the private cost of risk is £(106.25 − 105.66) million = £0.59 million in each case. For the two firms combined therefore the total risk-bearing costs will be 2 x 0.59 = £1.18 million. Spreading the risk over two firms has reduced total risk-bearing costs by nearly one half from £2.25 million to £1.18 million. The interested reader should soon confirm that a consortium of ten companies sharing equally in the costs and revenues of the project would yet further reduce the total cost of risk-bearing to £0.25 million.[9] It turns out that this tendency for risk-bearing costs to decline as they are spread over increasing numbers of firms is a special case of a more general theorem concerning risk-spreading − the Arrow−Lind Theorem − to which we turn in our next section.

## 7.4  Public Policy and the Discount Rate

In Section 7.2 the consumption and investment decisions of individuals and firms were discussed under the assumption of a perfectly operating market in state-contingent claims. If such a world existed the problem of the appropriate treatment of risk in public investments would hardly arise. The market prices of state-contingent claims, as with other prices in a first-best world, would suffice for the guidance of public investment decisions.[10] There is no case, as is sometimes argued, for the more lenient treatment of risk in the public sector on the grounds that the public sector is a more effective pooler of risks than the private sector, for, as Sandmo[11] points out, with properly functioning markets 'there are perfect opportunities for pooling of risks in the private sector, and the market risk margins represent a social evaluation of the risk associated with each type of investment'. Any argument for differing treatment of public and private investment is therefore dependent upon 'market failure' and is part of the theory of the second-best.

All market exchanges incur transactions costs and policing costs, but the market in insurance contracts or state-contingent claims faces particularly severe problems in these respects. It can be argued that private markets are not efficient because transactions costs inhibit exchanges of claims which would be mutually beneficial, or because some markets do not even exist because of the difficulty of policing agreeements. As already mentioned in Chapter 3 high policing costs stem from the existence of 'moral hazard'; that is, the possibility of a person buying a claim to resources contingent upon a given state of the world and then, by their actions, influencing the probability of that state of the world occurring. Celebrated examples occur in the fields of fire and motor insurance where the insured person may no longer take the care to avoid fire or accidents which he might otherwise have done. It should be emphasised that whether the existence of 'moral hazard' constitutes market failure is still a contentious issue. Transactions and policing costs exist in any resource-allocating system and cannot be assumed away. In the public sector there is no guarantee that administrators will behave as a benevolent despot would like. They may have their own reasons to be more risk-averse than is appropriate for society as a whole. The career prospects of an administrator who has been associated with several notable investment disasters are unlikely to be improved even if he correctly points out that he was simply implementing the appropriate investment criterion for the cases in point.

Ignoring these important problems of how decisions in the public sector will *in fact* be made it is still possible to ask whether *in principle* the public sector could improve on the allocation of resources to investment, especially under conditions of risk. Two possible lines of argument have been suggested. First, in the absence of perfect pooling possibilities in the private sector risk-aversion on the part of individuals may imply the rejection of projects which, if undertaken as part of a large public sector pool, could prove to be socially desirable. If the public sector is seen as undertaking a large number of statistically independent projects the 'Law of large numbers' will ensure that there is very little risk attached to the average returns per project in the total pool and that therefore the Government can adopt a risk-neutral attitude in project appraisal.[12] Second, and more recently, Arrow and Lind have shown that, under certain conditions, the social cost of risk associated with a given project declines to zero as it is spread over increasing numbers of people.[13]

The conditions attached to the Arrow–Lind theorem are clearly of central importance. The risk from a given investment must be capable of being spread over a very large population and, as with the 'pooling' argument, the returns to the project must be independent of the returns from existing investments. It is perhaps worth noting that in the simple example developed at the end of Section 7.3 this latter condition is fulfilled since there the returns to existing investments are assumed certain. In the context of resource allocation in the field of energy, however, the question arises as to how far the Arrow–Lind theorem is applicable. There are at least four important arguments to consider.

(i) If the Arrow–Lind case for ignoring risk in the public sector is to be accepted, mechanisms must be found which ensure the widest possible spreading of risk over the population. In their article Arrow and Lind envisage a government with a given budgetary target. An investment project brings to the Exchequer a flow of net revenue and the tax system is used to maintain the target. Thus if net returns are positive, taxes will be reduced, and vice versa.[14] Assuming that the taxes involved are broad-based and do not impinge on a small section of the population the effect is to spread the risk widely, as required.

Some risks, however, are of a type which, from both practical and theoretical standpoints, would appear very difficult to spread. The risk to life and limb of transporting highly flammable materials, for example, is not something which can be ignored in project appraisal. Those assuming such a risk may be able to be compensated, but it is not easy to see how the fiscal system could be used to effectively eliminate the risk from a social point of view.[15]

Further, even in simpler cases where the risk takes the form of differing possible revenue flows, the spreading mechanism relies upon compliant politicians willing to take the required fiscal action. If they decided that positive net revenues from a project provided an opportunity to increase public expenditure, rather than to decrease taxation, the nature of that expenditure would be of considerable importance. A decision to use net revenues to raise a particular social security benefit, for example, would clearly concentrate the risk of the project on a small group of people relative to the total taxpaying public.

(ii) Where the costs and benefits from a project take the form of 'public goods' or 'public bads' then any uncertainty concerning the magnitude of these costs and benefits clearly cannot be spread across the individuals affected. Since, by definition, all individuals experience the cost stemming from a 'public bad', all individuals will bear some risk if the magnitude of this cost is not certain. Possible environmental damages associated with energy projects, many of which have been discussed in Chapter 5 (p. 111), are clearly of relevance here. Examples would include the possibility of widespread atmospheric or water pollution stemming from various types of electricity generation.

It might be objected that much environmental damage is

of a local rather than national variety and that in principle there exists a possibility of transferring any risk from the local people to the nation as a whole. The administrative costs of such a scheme could be substantial, however, even were it conceptually possible, and these costs would have to be compared with the benefits from risk-spreading. Further, many readers may find that it stretches credulity too far to believe that the political system will, in fact, operate in this way. If it does not, then whatever may be the case in principle, in fact the risk will be born by the individuals directly affected.

(iii) In the field of energy there may also be some doubt as to how far the statistical independence assumption is likely to hold. Ultimately this question is an empirical matter and cannot be settled by *a priori* reasoning; however, the precise meaning of the independence condition is not obvious and a simple example may be of assistance. Table 7.1 contains entries for eight possible 'states of the world'. It is assumed that each state of the world is made up of a combination of possible events and that the probabilities attached to these events are objective and agreed. Thus the probability of new fossil fuel discoveries is assumed to be $\frac{1}{3}$ while the probability of technical advance is taken to be $\frac{3}{4}$ in the field of conventional fuel use and $\frac{1}{4}$ in the field of nuclear energy. The occurrence or non-occurrence of each of these three events permits the identification of eight possible 'states of the world' with a probability distribution given in column 5 of the table. Note that the probability of each event is assumed independent of the others. Thus the probability of 'state of the world' $A$, i.e. no new discoveries or technical advance, is simply $\frac{2}{3} \times \frac{1}{4} \times \frac{3}{4} = \frac{6}{48}$. Columns 6 and 7 of the table give the net returns to existing investments and to a proposed investment in nuclear energy respectively, for the various possible states of the world 'next period'. The precise magnitudes of these figures are of no significance and are intended merely for illustrative purposes. We have assumed, however, that net returns will be greater the more generally 'favourable' circumstances turn out to be.

Inspection of Table 7.1 reveals that the returns to the nuclear investment project are independent of existing returns in the following statistical sense. Whatever the outcome of

TABLE 7.1

| 'State of world' | New deposits of fossil fuel | Technical advance in fuel use | Technical advance in nuclear energy | Probability of 'state of world' | Net returns to existing investment | Net returns to nuclear investment |
|---|---|---|---|---|---|---|
| | | | | | $Y$ | $X$ |
| A | 0 | 0 | 0 | $\frac{6}{48}$ | 50 | 10 |
| B | 0 | 0 | 1 | $\frac{2}{48}$ | 50 | 20 |
| C | 1 | 0 | 0 | $\frac{3}{48}$ | 100 | 10 |
| D | 1 | 0 | 1 | $\frac{1}{48}$ | 100 | 20 |
| E | 0 | 1 | 0 | $\frac{1}{48}$ | 150 | 10 |
| F | 0 | 1 | 1 | $\frac{6}{48}$ | 150 | 20 |
| G | 1 | 1 | 0 | $\frac{9}{48}$ | 200 | 10 |
| H | 1 | 1 | 1 | $\frac{3}{48}$ | 200 | 20 |
| | | | | 1 | | |

Probability of new fossil-fuel discoveries = $\frac{1}{3}$.
Probability of technical advance in fuel use = $\frac{3}{4}$.
Probability of technical advance in nuclear energy = $\frac{1}{4}$.
0 designates non-occurrence of event. 1 designates occurrence.

existing investment, i.e. whether returns are 50 or 100 or 150, etc., the probability distribution of $X$ is unaffected. If $Y$ turns out to be 100, $X$ can take a value of 10 or 20 with the probability of the former being three times that of the latter. Similarly if $Y$ is 200, the distribution of $X$ is still 10 with probability $\frac{3}{4}$ and 20 with probability $\frac{1}{4}$. Knowledge of the $Y$ outcome in no way changes the probability distribution of the $X$ outcomes. The returns to existing investment are not related to the returns that may accrue from nuclear investment, and vice versa.[16] This result stems, of course, from the fact that in Table 7.1 the returns to nuclear investment depend entirely upon whether or not there is technical advance in the field of nuclear energy, whereas the returns to existing investments are not influenced at all by this event. In the case illustrated therefore the independence condition holds, and, if investment $X$ is to be undertaken in the public sector, evaluation should proceed on the basis of a riskless rate of interest.[17]

It is clear, however, that statistical independence of returns is not something that can be relied upon. If we assumed, for example, that the probability of technical advance in the field of nuclear energy was not independent of technical advance elsewhere or was related to fossil-fuel discoveries (perhaps because large finds could retard research and development) the entries in Table 7.1 would no longer be statistically independent. Similarly, if the returns from the investment in nuclear power depended not only upon technical factors in that industry, but also upon the occurrence or otherwise of the other events making up the state of the world, there would again be no reason to expect statistical independence to prevail. Thus we might expect the returns from an investment in nuclear energy to be greater in state of the world $B$ where conventional sources are failing, compared with state of the world $H$ where they are abundant.

Much hinges therefore on the assumptions made about the nature of public investments. If the statistical independence condition is considered plausible then the Arrow–Lind result follows, but if, following Sandmo,[18] the Government is seen as investing in many industries alongside the private sector then it can be shown that 'the public sector's discount rates should always contain a risk margin, and that this margin should correspond to the one used in the private sector for investment in the same risk class'. Where nationalised industries and private corporations are engaged in similar projects as, for example, in the oil industry, Sandmo's approach would appear to be more appropriate.

(iv) A final objection to risk-neutrality in the appraisal of public investments in the field of energy concerns the fact that some of them may involve 'irreversibilities'. Section 7.5 provides a brief review of this problem.

## 7.5 Irreversibility and 'Option Value'

Some decisions involve consequences which are very costly to reverse. In an extreme case it may even be technically impossible to revert to some preferred prior situation. These decisions are termed 'irreversible' decisions. A decision to develop the agricultural potential of parts of Florida by draining the Everglades, for example, would involve significant irreversibilities since, as Fisher and Krutilla[19] argue, a lower water table would irretrievably alter the aquatic plant and animal life of the area as well as the chemical constituents of the soil. Another celebrated example concerns the development of hydro-electric capacity at Hell's Canyon on the Snake River which would preclude a return to recreational use. Henry[20] cites as irreversible a hypothetical decision to demolish Notre-Dame cathedral in order to build a parking lot. This case illustrates the point that an important consideration is the 'authenticity' of any attempt to replicate the original position. Landscaping, for example, may be sufficient to restore an industrial or mining site for leisure purposes, but it will rarely be able to reintroduce the full range of flora and fauna which may be of great significance for the naturalist.

Many pollutants are degradable and may be rapidly assimilated by the appropriate environment. Others such as DDT, mercury or lead are nondegradable and may persist for many years. Thus a decision involving the disposal of these substances is essentially 'irreversible'. A particularly extreme case of this type is the problem of the disposal of radioactive wastes especially 'high-level' waste containing the 'actinides'.[21] These substances are lethal if inhaled in the minutest quantities and must be contained in the utmost security effectively for ever. Suggested methods of disposal themselves involve varying degrees of irreversibility. One possibility is that nuclear wastes could be buried in natural salt deposits which are geologically very stable formations free of circulating ground water. The problem is that once buried they rapidly become irretrievable and the option of using alternative containment methods in the future is lost.[22]

Recent work in the field of environmental economics has demonstrated that decisions which involve the loss of future 'options' have an economic cost which must be included in project appraisal. The flooding of an important geological formation may remove the option of using it in the future for different purposes, the decision to delay keeps all the options open. The existence of 'option value' wás first suggested by Weisbrod[23] and later developed by Cicchetti and Freeman.[24] As presented by the latter, 'option value' turns out to be what was termed the 'cost of risk-bearing' in Section 7.2, i.e. the amount that individuals would be prepared to pay to avoid uncertainty.

Imagine a person contemplating his demand for electricity in the next period. Suppose that there is sufficient capacity to supply some 'normal' load but that consideration is being given to increasing this capacity to deal with severe weather conditions. The individual knows that in the event of warm weather he would not demand electricity from these reserve sources, but that in cold weather his demand for electricity would increase. Let the probability of cold weather 'next period' be 0.5. In Figure 7.4 the two curves $U_1(Y)$ and $U_2(Y)$ represent the individual's utility of income functions for the cases in which electricity is available and electricity is unavailable respectively. In the absence of reserve plant the expected utility of this consumer will be $E(U) = 0.5\ \bar{U}_1 + 0.5\ \bar{U}_2$ assuming that his income level is $Y_0$.

Now imagine that the cold spell has arrived and that a monopoly supplier of electricity conducts an experiment. Specifically it asks the consumer how much he would be willing to pay to avoid a supply interruption. From Figure

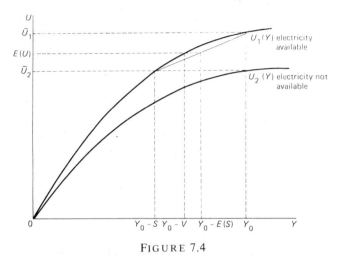

FIGURE 7.4

7.4 it is seen that the consumer would countenance a drop in income to $Y_0 - S$, at which point he would be indifferent between having extra electricity available and not having it available. The amount $S$ is therefore the consumer's surplus derived by the individual from consumption of extra electricity from reserve plant during the cold spell. Since the probability of the cold spell has been assumed to be 0.5 we deduce that the expected consumer's surplus from consumption in cold weather will be $E(S) = (0.5 \times S)$.

Suppose now that the monopoly supplier alters his experiment. Instead of waiting for the cold weather to occur and inquiring about consumer's surplus the electricity company asks how much the consumer would pay for the 'option' to purchase electricity at the prevailing price in the event of cold weather. From Figure 7.4 it is seen that the consumer would be willing to accept a reduction in income to $Y_0 - V$ if this ensured the availability of electricity in cold weather. Note that $V$, the maximum option price, is greater than $E(S)$, the expected consumer's surplus from extra electricity consumption in cold weather. Evidently the consumer is willing to pay more than the expected consumer's surplus for an assured supply. It is this excess of maximum option price $V$ over expected consumer's surplus $E(S)$ which Cicchetti and Freeman term 'true option value'.

Where individuals are unsure of what their demand for a commodity will turn out to be and where they are risk-averse, 'option value' will exist. In the above example the 'option' removed the risk that the consumer would demand peak electricity and find it unavailable, and this was shown to have a value quite separate from the expected consumer's surplus from the electricity itself. Because 'option value' is so closely related to the cost of risk-bearing, however, and because, as discussed in Section 7.4 there are circumstances in which this cost can be ignored in project evaluation in the public sector, it is reasonable to inquire whether, from a social perspective, 'option value' is important. It is in this context that the existence of 'irreversibilities' plays its part. In particular, Arrow and Fisher[25] have shown that, even where decision-makers are risk-neutral, if an investment project involves irreversible consequences there will exist 'a "quasi-option value" having an effect in the same direction as risk-aversion, namely, a reduction in net benefits from development'.

The existence of this 'quasi-option value' can be illustrated using a simple model very similar to that of Arrow and Fisher. Suppose that some energy project involves irreversible consequences, the flooding of land for hydro-electricity, the use of an area for strip mining, etc. Development involves both costs and benefits, but 'preservation' is assumed costless. It is assumed that there are two periods: 'now' (period 1) and the 'future' (period 2). Ignoring uncertainty for the time being it is seen that society has three choices: to develop now, to preserve now and develop later, or to preserve in both periods. The option to develop now and preserve later is not available because of the 'irreversibility'. In the event of immediate development total social benefit $S_d^*$ will be

$$S_d^* = (B_{d1} - C_{d1}) + B_{d2}, \qquad (7.10)$$

where    $B_{d1}$ = gross social benefit from development in period 1.

         $C_{d1}$ = costs of development in period 1.

         $B_{d2}$ = gross social benefit from development in period 2.

Note that for simplicity we have assumed that no costs are involved in period 2 if development net occurs in period 1. Also it has implicitly been assumed that society faces an all-or-nothing choice. 'Partial development' is not possible.[26]

If society chooses to preserve the site in period 1 there are two possible outcomes:

$$S_p^{**} = B_{p1} + B_{p2} \tag{7.11}$$

or $\quad\quad S_p^{***} = B_{p1} + (B_{d2} - C_{d2}), \tag{7.12}$

where $B_{p1}$ = gross social benefit from preservation in period 1, and the remaining notation can be read similarly.

In the absence of uncertainty there is no problem. Society presumably selects the option yielding the greatest total social benefit. Note that option (7.11) will be chosen only if $B_{p2} - (B_{d2} - C_{d2}) > 0$, i.e. if net social benefits from preservation in period 2 are positive. Similarly, option (7.12) will be chosen only if $(B_{d2} - C_{d2}) - B_{p2} > 0$. We can therefore rewrite equations (7.11) and (7.12) as a single expression representing total social benefits from options involving the preservation of the site in period 1:

$$S_p^* = B_{p1} + \text{Max}\ [B_{p2}, B_{d2} - C_{d2}].$$

A decision to develop the site in the first period will be made if $S_d^* > S_p^*$, i.e. if

$$B_{d1} - C_{d1} + B_{d2} - B_{p1} - \text{Max}\ [B_{p2}, B_{d2} - C_{d2}] > 0.$$

Let $(B_{d1} - C_{d1}) - B_{p1} = N_1$, where $N_1$ represents the net social benefits from development during period 1. The criterion then becomes

$$N_1 + B_{d2} - \text{Max}\ [B_{p2}, B_{d2} - C_{d2}] > 0. \tag{7.13}$$

The problem, of course, is that, by assumption, the magnitude of benefits and costs in period 2 is uncertain. With risk-neutral decision-makers the choice of whether or not to develop the site will depend on whether the *expected value* of expression (7.13) is greater or less than zero. The site will be developed if

$$N_1 + E(B_{d2}) - E\ [\text{Max}\ (B_{p2}, B_{d2} - C_{d2})] > 0. \tag{7.14}$$

It will be recalled from Section 7.2 that in the case of risk-neutrality the 'positive present certainty-equivalent value' criterion for project appraisal simply required *expected* returns to be discounted at the riskless rate of interest. In the presence of 'irreversibilities', however, it can be shown that using expected values of benefits and costs in (7.14) will not yield the correct expected value for the whole project. Replacing $B_{p2}$ and $B_{d2} - C_{d2}$ by their expectations we obtain

$$N_1 + E[B_{d2}] - E[\text{Max}(E\{B_{p2}\}, E\{B_{d2} - C_{d2}\})] > 0.$$

$$(7.15)$$

Suppose it happened, for example, that $E[B_{p2}] > E[B_{d2} - C_{d2}]$. In this case expression (7.15) would simply become

$$N_1 + E[B_{d2}] - E[B_{p2}] > 0. \qquad (7.16)$$

This is obviously no more than we might expect. Apart from the absence of discount factors, expression (7.16) is simply the usual 'positive expected present value' criterion for the development project. $N_1$ represents net social benefits from the project in period 1 and $E[B_{d2}] - E[B_{p2}]$ represents expected net social benefits in period 2.

Unfortunately, a moment's reflection will reveal that criterion (7.16) is not the same as criterion (7.14). Consider the expression Max $[B_{p2}, B_{d2} - C_{d2}]$. Clearly this must either take the value of $B_{p2}$ or, if not, it must be greater than $B_{p2}$. As long, therefore, as there is some positive probability that $B_{d2} - C_{d2} > B_{p2}$ it must be concluded that $E[\text{Max}\{B_{p2}, B_{d2} - C_{d2}\}] > E[B_{p2}]$. It follows immediately that (7.16) overestimates the true expected net social benefit from the development project and that adopting a criterion such as (7.16) will be unduly favourable to the undertaking of projects with irreversible effects.[27]

## 7.6 Problems of Using Probabilities

Although somewhat abstract, the analysis presented so far has permitted the identification of some important factors bearing upon the direction of any adjustment for risk in project selection in the public sector. In particular three major factors stand out.

(i) The degree to which an investment project produces 'public goods' or 'public bads'. The risk associated with variations in these cannot be spread over the population as the Arrow–Lind Theorem requires.

(ii) The extent to which returns are statistically dependent or independent of returns from other investments.

(iii) Whether a project involves effects which are 'irreversible'. We have argued that the appraisal of some types of investment in the field of energy may involve adjustments for uncertainty under each of these headings, and that discounting expected returns, at a riskless rate of interest will not yield the correct certainty-equivalent present value.

These observations still leave the rather substantial problem of deciding on the *quantitative* size of any reduction in net social benefits from a project due to risk. If statistical dependence of the returns with other investments is the only problem, and if existing financial markets are considered sufficiently developed to permit the widespread, if not perfect, pooling of risk in the private sector, one solution in principle is to observe what adjustments are made in the private sector on investments in the same 'risk-class'. Where, however, arrangements for risk-pooling in the private sector are undeveloped the problem is less tractable. 'Risk margins' for a given class of investment will vary across individuals because the necessary markets do not exist to permit the exchanges necessary to bring them into equality.[28]

Intuitively it seems plausible that, in the absence of a single market-determined adjustment for risk, some 'average' of those used in the private sector might be appropriate. It might even be surmised that the more any particular individual is affected by a project the greater should be the weight given to his preferences concerning risk or his 'risk margin' on the class of investment under consideration. Under certain conditions these suggestions can be demonstrated rigorously,[29] but the informational requirements necessary to carry through such an analysis are obviously immense. They are, however, no more startling than the informational requirements theoretically necessary to determine the optimal provision of pure public goods, and it is clear that in both cases political institutions will play a large part in determining outcomes in practice.

In addition to the problem of specifying the quantitative importance of the results obtained in the foregoing sections, it is possible to cast doubts on the validity of the whole approach. In particular, much of the analysis is dependent upon the existence of an objective probability distribution for states of the world. Such information is rarely available. Formally, the consumption and investment decision can be analysed using 'subjective probability estimates' which may differ between people. But in the sphere of public policy this gives rise to trouble. In the absence of objective data, whose probability estimates are to be used? Much disagreement over policy issues has its roots in differing probability estimates, and statements about these estimates – for example concerning the safety of various forms of electricity generation, or the likelihood of extensive new discoveries of fossil fuel – are usually extremely contentious.

· This problem is obviously most acute when the concept of probability as 'relative frequency' is most difficult to apply. Unfortunately this is clearly very often the case in the sphere of energy. The probability that the accumulation of carbon dioxide will result in serious climatic changes is not something which can be assessed from repeated experiments. The probability of a major disaster at a nuclear power station may be estimated using 'safety analysis', but from a relative-frequency point of view all that is known is that such an accident is yet to occur.[30] As Hirshleifer comments, the state-preference approach provides a theory which contains uncertainty about future states of the world but it 'does not contain the "vagueness" we usually find psychologically associated with uncertainty'.[31] In the following section some alternative approaches to choice under uncertainty are outlined, approaches which do not rely on the existence of objective probability distributions over states of the world.

## 7.7 Decision Theory

In the absence of objective knowledge concerning probabilities a number of decision rules have been formulated by economists. The most celebrated of these approaches are illustrated below, using a very simple example.[32] Suppose that a decision-maker is to choose between three mutually exclusive investment strategies. The pay-off associated with each strategy is dependent upon which of three mutually exclusive states of the world occurs. Thus in the matrix below, strategy 1 produces a pay-off of 200 in state of the world *a*, strategy 3 a pay-off of 150 in state of the world *c*, etc.

*Pay-off Matrix*

### STATE OF THE WORLD

|  |  | *a* | *b* | *c* |
|---|---|---|---|---|
| *Investment* | 1 | 200 | 0 | 50 |
| *strategy* | 2 | 20 | 30 | 40 |
|  | 3 | 15 | 90 | 150 |

### (*a*) The Maximin Criterion

Faced with these prospects a very conservative strategist might assume that the worst will happen whichever strategy he chooses and that consequently he should pick the alternative which offers the largest pay-off in the least favourable state of the world. Making a list of the minimum pay-off for each strategy the decision-maker would obtain the following result.

### MINIMUM PAY-OFF

|  |  |  |
|---|---|---|
| *Investment* | 1 | 0 |
| *strategy* | 2 | 20 |
|  | 3 | 15 |

Clearly in this case strategy 2 offers the 'highest safety net'. Whatever happens, if strategy 2 is chosen, the pay-off cannot fall below 20.

One obvious criticism of the maximin criterion is that it concentrates entirely on the worst possible outcome for each strategy and ignores all the other possibilities. Looking at the entire matrix of pay-offs a different decision-maker might note that by choosing strategy 2 the option of some very high returns (200 in the case of strategy 1 and 150 in the case of strategy 3) is ruled out. The *worst* that can happen if strategy 2 is chosen may be somewhat higher than is the case for the other strategies, but the *best* it can do is considerably less than is the case for the others.

### (b) *The Minimax Regret Criterion*

Our new decision-maker may reformulate his choice criterion along the following lines. Consider strategy 2. If state of the world *a* occurs the payoff is 20 and this more enterprising decision-maker is sure to note how much better it would have been in this case to choose strategy 1. Similarly, if state of the world *b* occurs, the decision-maker will reflect ruefully on the pay-off he might have received had he chosen strategy 3. In general, instead of considering the actual pay-offs accruing to a particular strategy, the decision-maker considers the 'regret' that is experienced at not having chosen some best-alternative strategy. 'Regret' in this context is simply the shortfall of the actual pay-off from the best alternative, given the state of the world. Thus, if strategy 2 is chosen and state of the world *c* occurs the regret will be $150 - 40 = 110$. The full matrix of regrets is given below.

### STATE OF THE WORLD

|  |  | *a* | *b* | *c* |
|---|---|---|---|---|
| *Investment* | 1 | 0 | 90 | 100 |
| *strategy* | 2 | 180 | 60 | 110 |
|  | 3 | 185 | 0 | 0 |

From this matrix of figures it is seen that the maximum regret associated with investment strategy 1 is considerably lower than the maximum regret associated with the two alternative strategies. Thus if the decision-maker wishes to minimise his potential regret he will choose strategy 1.

It remains true, however, that attention is still focused on a single outcome for each strategy instead of all possible outcomes. Thus although the two criteria so far described have opted for strategies 2 and 1 in turn, it is evident from the regret matrix that strategy 3 does have some advantages. In particular it is the favoured strategy in terms of having the best pay-off and hence no regret for two out of the three possible states of the world. However, without some knowledge of probabilities it is difficult to derive choice criteria which take account of all possible outcomes.

## The Bayes Criterion

One set of subjective probabilities for states of the world of which decision theory has made use is the set of *equal* probabilities. In a state of complete ignorance, it might be argued, we might as well assume that all states are equally probable and work out the expected returns on this basis. This type of approach has been graced with the term 'the principle of insufficient reason'. It enables us to calculate the strategy with the greatest pay-off 'on the average' assuming equally probable states of the world. The expected pay-off of each strategy $E(P)$ is given in the following table.

| Investment strategy | | Expected pay-off |
|---|---|---|
| 1 | $(\frac{1}{3} \times 200) + (\frac{1}{3} \times 50)$ | $= 83\frac{1}{3}$ |
| 2 | $(\frac{1}{3} \times 20) + (\frac{1}{3} \times 30) + (\frac{1}{3} \times 40)$ | $= 30$ |
| 3 | $(\frac{1}{3} \times 15) + (\frac{1}{3} \times 90) + (\frac{1}{3} \times 150)$ | $= 85$ |

This criterion recommends strategy 3 as the best alternative with an 'expected' pay-off of 85.

From this cursory review of some of the major approaches to decision theory it will be obvious that none provides a clearly superior choice criterion to each of the others. Everything will depend on the individual decision-maker's attitude to uncertainty and it has been shown that any one of the three investment strategies investigated might be justified on quite 'reasonable' grounds.

## 7.8   Implications for Policy Formulation

From the earlier sections in this chapter it is evident that any policy towards uncertainty requires solutions to three major problems.

(i) Attempts must be made to define the 'states of the world' which are possible and the outcomes or 'pay-offs' which these states imply. Each section from 7.2 to 7.7 has depended upon explicit knowledge of the possible outcomes involved. In reality the collection of this type of information may involve substantial costs and require extensive research. The environmental effects of various pollutants, for example, are frequently very difficult to estimate especially when 'synergistic effects' are involved. Some toxic substances may vary considerably in the danger they represent, depending on the presence or absence of yet other substances. The enormous possible number of combinations of pollutants which may result from modern industrial activity clearly represents an information problem of great complexity.

(ii) Probability estimates for the various 'states of the world' are required. In Section 7.6 we have already drawn attention to some of the problems involved. A fundamental decision must be made as to whether, in the absence of objective probabilities, some subjective estimates are permissible, and if so, who is to make them. The brief outline of attempts to devise decision rules ignoring probabilities given in the last section revealed their considerable shortcomings. Concentrating on one state of the world only and ignoring for the purposes of decision-making all other possible outcomes, or alternatively arbitrarily assuming all states of the world have an equal probability of occurring; these approaches it can be argued are unduly restrictive. In a powerful appeal for the use of subjective probability estimates in decision-making, for example, the Committee on the Principles of Decision Making for Regulating Chemicals in the Environment argue as follows. 'Probability theory provides an unambiguous and logically consistent way to reason about uncertainty. In fact, a forceful argument can be made that any logical process for reasoning about uncertainty is equivalent to probability theory'.[33]

Once it is accepted that probability estimates are a useful tool for decision-making, even when information on relative frequency is unavailable, the difficulty remains of deciding how they are to be derived. Clearly scientific experiments in the laboratory or statistical evidence from the past may provide some useful guidance. 'Hazard analysis', which attempts to estimate the dangers involved in using various chemicals (usually by experiments on animals, or by observing populations already exposed to differing concentrations), is very important for improving information about health costs.[34] The judgement of engineers and scientists conversant with the theoretical and practical problems involved would be required to derive probability estimates for the financial costs involved in developing new sources of energy such as wind, wave and solar power. Past experience and the opinions of qualified geologists might be sought when assessing the prospects of discovering deposits of fossil fuel in a particular area. The construction of models and prototypes of reactors, boilers, windmills, etc., can be regarded as investment in information concerning the probable performance and cost of the full-scale item.[35] Thus, if subjective probabilities are to be used they are usually seen as coming from 'experts'. As the Committee cited earlier explains, 'the probability, for example, that continued use of an aerosol propellant will affect the ozone layer must be taken as scientific judgement: nothing more or nothing less. The probability value acts as a summary of the scientific judgement of experts in a form useful to the policy maker. . . . '[36]

At least two objections are immediately apparent to the use of 'experts'. First, 'experts' do not always agree with one another, nor are they immune from the enthusiasms of less-expert individuals. Fluidised-bed boilers for the combustion of coal, tidal barrages, windmills, various types of nuclear reactor, all have their supporters and sceptics, and there would seem little reason to suppose that agreement could always, or even usually, be reached on probability estimates for various outcomes. Second, even where the 'experts' agree with each other, awkward ethical questions are raised in some circumstances if large sections of the population persist in disagreeing with them. The fundamental ethical postulate of

Paretian welfare economics that 'individual valuations count' would be violated if probability estimates were imposed by experts. As we saw in Section 7.2 (p. 192), state probabilities enter into individual utility functions and thereby determine the certainty-equivalent value represented by a given state-contingent flow of returns. It is the aggregate certainty-equivalent value over all individuals affected by an investment project that is relevant in determining whether or not it should go ahead.

Perhaps the strongest defence of including subjective probability estimates in project appraisal is similar to the familiar defence of attempting to value lives saved, noise pollution and other 'intangibles' in cost—benefit studies. Including such estimates forces the planners and politicians to be *explicit* about the assumptions upon which their decisions are based. It makes it easier for inconsistencies to be exposed in the assumptions underlying the evaluation of different projects, and it opens the way for 'sensitivity analysis', that is it enables us to ask the question 'what set of probabilities over states of the world would have the effect of reversing our provisional result?'

(iii) The third major problem area, and the one to which we have devoted the most attention, is that of deciding on the appropriate adjustments to the *expected returns* which will yield the certainty-equivalent value of the project. It has been noted in earlier sections that in the presence of perfect markets in state-contingent claims, or where risks are spreadable and where the returns from á particular project are statistically independent of returns elsewhere, fairly clear results can be derived to guide public policy. Where markets are imperfect, risks not spreadable and returns not statistically independent, however, simple exhortations to imitate the private sector or to use a riskless rate of discount are no longer appropriate.

In principle, detailed knowledge of the effects of a project on each individual over all states of the world along with information about the preferences of all individuals affected, would be required to calculate the present certainty-equivalent value in these circumstances. Clearly some 'rules of thumb' are necessary if such enormous informational requirements are not to inhibit completely attempts at project appraisal. The most common suggestion is that the rate of interest used to discount expected net benefits should be raised, the precise extent of this increase being dependent upon an assessment of the 'riskiness' of the project. This in turn would depend upon the 'publicness' of the benefits and costs, the degree of risk-aversion experienced by the affected population, the presence or otherwise of 'irreversibilities' and so forth. A major theoretical difficulty here, however, is that increasing the rate of discount will discriminate against long-term projects. Considerations of time-preference and risk-preference are being mixed up together when they are in principle quite separate. In the simple two-period analysis presented earlier it was possible to derive a single 'risky rate of interest'. In a multiple-period analysis there will theoretically exist a separate 'risky rate of interest' applicable to the discounting of each period's returns. Even if it is assumed, as is usual in project appraisal, that the pure time-preference rate remains constant from period to period, there is no reason to expect a constant increment in this rate to allow adequately for the risk associated with the returns in each period.

An alternative possibility is to discount expected costs and benefits from a project at a riskless rate but to adjust the present values of these streams, again according to a rough assessment of the importance of the elements discussed earlier. This has the advantage that the two procedures – adjusting for time-preference and risk – are kept separate. It must be frankly admitted, however, that upward adjustments in the present value of cost streams and the reverse for benefit streams, or the use of lower discount rates for calculating present values of cost streams and higher ones for the evaluation of benefit streams, both appear somewhat arbitrary when considered alongside the theoretical information requirements.

In Chapter 5 (p. 129) it was noted that lack of information concerning damage-cost functions created similar problems for achieving an optimal allocation of resources in the field of environmental pollution. Very frequently the political system lays down 'standards' which are deemed 'acceptable' for the time being but which may not be 'optimal'. Examples include vehicle exhaust emission standards and standards regulating water quality.[37] It is perhaps not surprising that in the absence of sufficient information about individual preferences the problem of uncertainty can be tackled along similar lines. Instead of attempting to calculate certainty-equivalent values for varying profiles of state-contingent returns there may be circumstances in which it is expedient for the political system to lay down safety standards. These standards would take the form of requiring that the probability of certain states of the world occurring does not exceed a given level, or alternatively that the costs associated with a given state of the world are kept within stipulated bounds. No investment project with state-contingent returns or state probabilities outside these limits would be approved.

A good example of this procedure occurs in the field of nuclear power. The design of reactors in the United Kingdom is undertaken with specific safety standards in mind. Figure 7.5 illustrates the 'limit line' which plots the maximum permitted probability per year for a single reactor of various accidental releases of radioactivity. In particular the criterion suggested by F. R. Farmer and taken up by the Atomic Energy Authority is that there should be a probability of less than one in a million per year of a release of one million curies of iodine −131.[38] This criterion in itself will not enable most people to judge how 'acceptable' or otherwise it

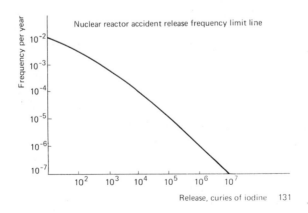

FIGURE 7.5

may be and this type of information is usually put in a form nearer to everyday experience, or at least imagination, by comparing it with the probable number of yearly fatalities from air crashes, fires and dam failures. However, it is worth repeating that both for estimates of the likely consequences of various accidents and for estimates of the probabilities of those accidents great reliance is placed inevitably on the judgement of experts and the quality of their scientific and safety analysis.

## 7.9   Conclusion

There is no single method of coping with uncertainty which commands universal support. For 'commercial risks' – that is, spreadable risks involving cost and price changes – technological development and changes in demand conditions, the approach of identifying 'risk-classes' and using a risk margin on the discount factor equivalent to that used on other similar investments would seem appropriate. This approach does depend, however, on the assumption of reasonable opportunities for risk-pooling in the private sector and on good information on state probabilities. Where the statistical independence assumption is considered plausible the Arrow–Lind theorem might be invoked and a riskless rate of discount applied to expected returns. The theorem holds even in the absence of efficient markets in state-contingent claims, but in Section 7.4 (p. 210) we expressed doubts as to whether the other necessary conditions would hold in the field of energy policy.

For risks involving public goods and bads, environmental effects and irreversibilities some qualitative results are derivable from the literature. In particular, there is no question of the Arrow–Lind result applying in these circumstances, and some adjustment for the social risks involved must be made. Unfortunately, precise policy prescriptions founder on the problem of obtaining information about probabilities, states of the world and individual preferences. Decision rules requiring somewhat less information about state probabilities and assuming particular attitudes to uncertainty on the part of decision makers were outlined in Section 7.7. The U.K. Government appear to have had some such criterion in mind when supporting courses of action which 'may be described as "robust" or as producing "minimum regret" ' in the context of long-term energy strategy.[39]

Environmental and health risks are most likely to be tackled by the setting of 'politically acceptable' standards along the lines of those discussed in Chapter 5 (p. 129). These may involve limiting the discharge of various pollutants via taxes or regulatory instruments, or in cases where the discharges themselves are uncertain, for example in the transport of oil by sea or the generation of electricity from nuclear power, insisting on safety standards for the equipment and plant involved.

# 8. ENERGY ANALYSIS

## 8.1 Introduction

The subject-matter of this chapter differs from that of others in this book in that it is only indirectly concerned with the economics of energy. This chapter is concerned with an analytical technique which is variously known as energy analysis, energy accounting, energy budgeting, or energy costing. The aim of energy analysis is to calculate the energy inputs required for any project, which could be anything from the development of oil shale resources, the construction and lifetime operation of a nuclear power station to the expansion of agricultural output.[1] In this analysis the £ or dollar *numéraire* of economic analysis is replaced by an energy measure *numéraire*, such as joules, kilocalories or Btus (British thermal units). A variant of energy analysis, which is of special relevance to the general argument of this book, is known as net energy analysis. Net energy has been defined as the amount of energy that remains for consumer use after the energy costs of finding, producing, upgrading and delivering the energy have been paid.[2]

Energy analysts hold a variety of views on the role and uses of energy analysis. Some regard it simply as a technique for generating information which can be used to improve policy decisions. For these analysts the technique has (correctly in our view) no normative significance. Others, however, are quite explicit in claiming a normative significance for the technique. They regard it as an instrument which should be used in the evaluation of policy proposals. The U.S. Congress has endorsed this view of net energy analysis in Public Law 93–577. Section 5(*a*) 5 of the Non-Nuclear Energy Research and Development Act of 1974 states that 'the potential for production of net energy by the proposed technology at the stage of commercial application should be analysed and considered in evaluating proposals'.[3] The U.S. Energy Research and Development Administration has stated that it plans to integrate the net energy analysis of alternative technologies into a national plan for setting research and development priorities.

Since, at least in the United States, energy analysis has been accepted by some decision-takers as having an evaluative function, it seems to us important that the uses and limitations of this technique should be widely appreciated. In this chapter special consideration will be given to some differences between economic and energy analysis. This is because some energy analysts have claimed that energy analysis provides superior information to policy-makers than that which is supplied by economic analysis.[4]

Apart from this introduction this chapter has six sections. The various methodologies of energy analysis are considered in Section 8.2, while the problem of the choice of *numéraire* is considered in Section 8.3. The general issue of the use of energy analysis in the formulation of energy policy is considered in Section 8.4. Sections 8.5 and 8.6 consider, respectively, some matters relating to energy analysis and the goal of energy conservation, and the use of energy analysis in the determination of the 'desired' inter-temporal allocation of resources. Finally some conclusions are presented in Section 8.7.

## 8.2 Methodology

It is important to understand that, as generally practised, energy analysis is concerned with those forms of energy which exist as a stock rather than as a flow resource. Thus it is basically concerned with fossil-fuels, and to a lesser extent with uranium.

Energy analysis has been described as a 'systematic way of tracing the flows of energy through an industrial system so as to apportion a fraction of the primary energy input to the system to each of the goods and services which are an output of the system. Thus the result of an energy analysis is a set of energy requirements for all the commodities produced'.[5] The basic idea of energy analysis can be illustrated as follows.

Consider a product $X$, the production of which will require both direct and indirect energy inputs. Assume that the only direct energy input is coal, which will be removed from the energy stock with an energy cost of $E_X$. The mining of the coal and its transport to the producer of $X$ will have involved the use of energy. Similarly the extraction of other raw materials (which are treated as being free gifts of nature), their processing and transport to the producer of $X$ will have involved energy costs. This chain of energy costs can be represented as $E_{M,1} + \cdots + E_{M,N}$. Capital is made up of both energy and other inputs, and thus there will be a chain of energy costs, $E_{K,1} + \cdots + E_{K,M}$, describing the production of capital in energy terms. This chain must be annuitised to allow for the durability of capital. (Notice that for questions of the type, how much energy would be required to produce another unit of $X$? the energy embodied in capital is irrelevant. Many energy studies, however, have overlooked this fact.) Labour is sometimes treated like other inputs and given an energy value, $E_{L,1} + \cdots + E_{L,P}$, but in other cases it is excluded from the analysis. The former case gives a single-factor model of the economy, while in the latter case there is a two-factor model. The total energy cost of producing $X$ is then

$$E_X + (E_{M,1} + \cdots + E_{M,N}) + (E_{K,1} + \cdots + E_{K,M})$$
$$+ (E_{L,1} + \cdots + E_{L,P}).$$

Before we consider some of the analytical problems involved with energy analysis, such as the allocation of joint energy costs and the determination of the boundary which will separate those energy inputs which are to be counted in the analysis from those which are to be excluded, we will consider briefly the three principal methods which have been used in the evaluation of energy requirements.[6] These are:

(1) Process analysis
(2) Statistical analysis
(3) Input–Output analysis.

Process analysis involves the identification of the network of processes involved in the production of a good, the analysis of each process to determine its set of inputs, and the assignment of an energy value to each input. The aggregation of these energy costs gives the total energy cost of the good in question.

Statistical analysis uses data from published statistics, such as the U.K. Census of Production to identify all the energy inputs to an industry. These data are then used to calculate the energy cost per unit of output. The use of national statistics implicitly defines the system boundary, and energy analysts have put forward a number of methods for dealing with inputs which cross this boundary.

The input–output approach involves taking an input–output table and using it to calculate the list and quantity of commodities required to produce any given commodity. Since the information in the input–output table is in monetary terms (the $a_{ij}$th entry shows the amount of commodity $i$, in £, required as a direct input to produce £1 worth of commodity $j$) it is not in a form which is immediately applicable to energy analysis. For this purpose the monetary terms must be converted into physical terms so that the tables can be used to indicate the amount of primary energy per unit of output.[7] With the input–output method it is important to note that the calculated energy costs are based on financial data, and that imports are usually treated as having the same energy requirements as if they were produced domestically (and thus that the technology at home and abroad is the same).

The use of these different methods often leads to substantially different estimates of the energy costs of particular products. Thus Chapman[8] has reported that the use of the U.K. input–output table gave the energy required to produce 1 tonne of aluminium as 60 GJ, while a number of process studies gave results in the range 260 to 290 GJ. There are a number of reasons why different methods of analysis can give rise to such different results. These include the fact that both the allocation of energy costs and the choice of sub-system (to determine which inputs are to be counted) involve arbitrary decisions.

Energy inputs are often joint to a number of different products. As with financial accounting, energy analysts have used a variety of conventions for the allocation of these joint

energy costs to specific outputs, such as allocation according
to weights, prices or enthalpies[9] of the different products.
The problem of joint energy cost allocation is very important
in some net energy analysis studies; for example, those of
nuclear reactors. Uranium is often a joint product, as with
gold in South Africa and phosphates in Florida. Economists
have long recognised that joint costs can only be allocated in
arbitrary ways. While recognising the truth of this, some
energy analysts have argued that the cost-allocation conventions should be chosen according to the aims of the study,
and by doing this they claim that the allocation ceases to
be arbitrary.[10] Others, however, have concluded that the
inability of energy analysis and net energy analysis to solve
the joint cost allocation problem means that major energy
systems – such as nuclear reactors – cannot be realistically
analysed using these techniques.[11] This is a position which
we endorse.

A related point concerns the choice of the sub-system.
Since energy inputs can be traced through an economic
system virtually without end, costing requires the choice of a
cut-off point. This choice is largely arbitrary and the choice
of different cut-off points will result in different assess-
ments.[12] A similar point relates to the choice of the spatial
boundary and to the energy costing of imports. Various
conventions have been suggested: such as attaching a zero
cost on the basis that the imports are not produced in the
home country, attaching a cost reflecting the energy content
of 'average' exports of the same monetary value, and attaching
an energy cost equal to that which would be incurred if the
same product was produced domestically. It should be noted
that while this lack of an agreed costing procedure does not,
in our view, constitute a fatal criticism of the methodology
of energy analysis and net energy analysis it does impose a
severe limitation on its use as a practical tool for decision-
taking.


## 8.3  Choice of *Numéraire*

In both economic and energy analysis the problem is encountered of how to add together different products. The problem is solved in economics by using a *numéraire* defined in terms of money, such as £ or dollars. In energy analysis, as we have already noticed, the usual practice is to use an energy unit of account, such as joules.

For energy analysis, however, it has been suggested that the adding-up problem can be avoided by recording all energy flows separately.[13] Thus there would be separate flows of coal, oil, gas, etc., and each type of energy could be sub-divided by a quality grade. The essential problem with this procedure is clearly that it would make different commodities and different technologies, using different mixes of energy, non-comparable. How would a product $X$ using 2 kilos of coal and 1 gallon of oil be compared with a product $Y$ using 10 therms of natural gas and 20 kWh of electricity? Recognising this problem, energy analysis studies generally use an energy unit of account.

If different types of energy are to be added together it is necessary that they be normalised for quality. This may be done by defining quality in terms of ability to do work. A calorie of electricity is able to produce more work than a calorie of coal or oil, which in turn is able to produce more work than a calorie of sunlight.[14] Higher-quality forms of energy are able to undertake work that is impossible for lower-quality forms; for example, electronic communication is impossible without electricity. Energy quality ratios have been determined approximately for some forms of energy in a study at the University of Florida, and are given in Table 8.1.

These ratios suggest that 1 unit of coal can do the same work as 2000 energy units of sunshine. 'Thus, if we want to compare a solar collector for heating water with a coal-fired water heater . . . we can make the energy sources equivalent by the use'[15] of the data in Table 8.1. It is important to notice that these ratios do not distinguish those energy resources which exist as a flow (e.g. solar energy) from those which exist as a stock (e.g. coal). Thus, and this is something to which we will return, the energy quality *numéraire*, as we would expect, makes no reference to the *relative scarcity* of different fuels. In the example quoted it is explicitly stated that solar energy

TABLE 8.1

*Energy Quality Factors*[a]

|                       | Sun  | Wood  | Coal   | Electrical |
|-----------------------|------|-------|--------|------------|
| Solar equivalents     | 1    | 0.001 | 0.0005 | 0.00014    |
| Sugar equivalents     | 100  | 0.01  | 0.005  | 0.0014     |
| Wood equivalents      | 1000 | 1     | 0.5    | 0.14       |
| Coal equivalents      | 2000 | 2     | 1      | 0.28       |
| Electrical equivalents| 7200 | 7.2   | 3.6    | 1          |

[a] These are preliminary values and are expected to change as improved data become available.

SOURCE: C. D. Kylstra, 'Energy Analysis as a Common Basis for Optimally combining Man's Activities and Nature's, chap. 17 in G. F. Rohrlich (ed.), *Environmental Management* (New York, Ballinger, 1976) p. 274.

and coal, once they are made comparable in quality terms, are equivalent.

Some of the problems involved in using a common energy unit of account can be illustrated by a simple example which contrasts such a physical measure with the economist's concept of opportunity cost.[16] Assume the existence of some resource – call it 'oil' – which is homogeneous and in finite supply. In addition assume that all the deposits of this resource are equally accessible, that technology is fixed, and thus that there is no change over time in the physical inputs required to extract 1 ton of this resource. Thus each ton can be obtained at a constant expenditure of energy. Finally, assume that there are no substitutes available for this resource.

As the physical exhaustion of this resource approaches its energy cost (in joules), as measured in energy analysis calculations, will remain constant. In contrast economic analysis, using the £ *numéraire*, would show the price of 'oil' increasing to reflect its increasing scarcity. This would probably happen in two ways. Suppliers, seeing the coming exhaustion of their product and knowing of the absence of the possibility of substitution, would raise its price. Second, if there exists a futures market, dealers in this market would offer higher prices for it as the time of its physical exhaustion approached.

In the absence of a futures market the price will still rise because of the reaction of suppliers. The essential point is very simple; using economic analysis and the £ *numéraire* the price of this resource will rise over time to reflect increasing scarcity, whereas its energy cost would remain constant.

It is important to understand that in making this point we are not claiming that market-determined prices for exhaustible resource are in some sense 'correct' (see Chapter 3, p. 66). Such prices may, for example, give too little weight to the interests of future generations. However, we do believe that economic analysis has a fundamental advantage over energy analysis in its recognition of the importance of relative scarcity.

The example just considered is important in terms of the *raison d'être* of both energy and net energy analysis. Thus it has been suggested that:

there has been a growing realisation that the financial costs of materials and products do not provide an adequate description of the resources needed for their production. When there are no shortages of any inputs to the production system financial analysis provides a convenient decision-making framework. However, if one input does become scarce, then the implicit assumption of substitutability, inherent in financial systems analysis, leads to false conclusions. For a wide variety of reasons a number of investigators have focused their attention on the physical inputs, such as tons of steel and kWh of electricity, needed to make particular products. The forecasts of energy shortages coupled with the realisation that energy is an essential input to *all* production processes have concentrated attention on the inputs to, or energy cost of, various products.[17]

If financial analysis is interpreted to mean economic analysis using market price data, then this quotation would seem to imply that, because of deficiencies in the market mechanism, economic analysis will either fail to identify changes in relative prices over time due to changes in relative scarcities, or else it will do so later than energy analysis.[18] Now while we would certainly not claim that markets operate perfectly, we fail to see how energy analysis can improve our knowledge of the situation.

Because of the importance of this point it is worth noting that it has been claimed that an advantage of energy analysis over economic analysis is that in the absence of technological change net energy estimates will not change with changing dollar values.[19] This is a good example of the kind of mis-understandings of the workings of the economy that can occur by viewing all inputs in terms of transformed energy. Clearly for constant net energy estimates it is not sufficient merely to hold technology constant. Net energy estimates depend not only on the state of technology but also on the structure of industry and the set of relative prices when they are made. A change in the structure of relative prices will alter the choice of inputs (factor proportions) and their total energy content. Changes in relative prices will lead to substitution effects even in the absence of changes in technology.

The valuation of fuels in terms of their relative heat content implicitly assumes that this is their only attribute which consumers value. But this is not the case.[20] Consider a producer's choice between several different types of energy inputs, such as the generation of electricity from coal, oil or nuclear power. In economic analysis the inputs will be allocated efficiently when the marginal product of each unit in each output is the same. That is when the marginal rate of transformation is common between all units.

Different types of energy may be used in various ways. Thus coal can be used directly, or it can be used indirectly by being converted into electricity. If the relevant objective is an efficient allocation of resources and the necessary set of conditions for this are satisfied, then one economic measure of how one fuel type should be converted into another is given by the marginal rate of transformation of factors in production.[21] In cost terms this means that the various fuel inputs should be valued at their marginal production costs. Where fuels are purchased by consumers the conversion factor is given by the consumer's marginal rate of substitution. If the conditions for an efficient allocation of resources are satisfied this will be measured by the marginal cost of fuel to the consumer.

What these concepts tell us is that in both production and consumption a calorie is not necessarily a calorie. Fuels have a number of attributes and heat content is but one. Two fuels with the same heat content but other different attributes (in terms of cleanliness, storability, etc.) would have different marginal costs. In terms of economics the fact that two forms of fuel have the same heat content does not make those fuels identical. But in energy analysis the use of the energy unit of account obscures this important difference. In our view the calculation of energy costs using the energy *numéraire* makes energy analysis irrelevant to the process of resource allocation, both intra- and inter-temporally. If, however, the use of this *numéraire* is dropped then energy analysis has no foundation.

In view of the fact that, for example, the U.S. Congress has mandated the use of net energy analysis, this argument that energy analysis has no relevance to resource-allocation decisions is of extreme importance. It precludes energy analysis from being used for many of the purposes claimed by energy analysts. For example it has been said that if the thermal efficiency of power stations is only 25 per cent then 75 per cent of the energy input is lost.[22] But within the economic system consumers are expressing a preference for a secondary fuel input over a primary fuel input. The price which they pay for electricity will reflect the opportunity costs of all the inputs, including coal, used to make it. This is true irrespective of how much of the potential energy in coal would be used in its alternative uses. Unless consumer preferences are not to count it is misleading to talk of the therms not directly converted into electricity as being lost.

## 8.4 Energy Analysis and Energy Policy

Most energy analysts are quite explicit in claiming that energy analysis is not an energy theory of value. However, this claim has been made[23] and further it has been suggested that allocative decisions should be assessed using both monetary and energy units of account.[24] It is therefore pertinent to inquire into the circumstances under which energy analysis and economic analysis would give rise to the same allocations of resources.[25]

For simplicity we assume a closed economy in which all markets are perfectly competitive; that all producers are profit-maximisers and consumers are utility-maximisers. It is assumed that all production and utility functions are concave to the origin, continuous and twice differentiable. It is also assumed that a single homogeneous commodity 'energy', which is represented as $Q$ and measured in kilocalories, has the production function

$$Q = f(X_1, \ldots, X_n),$$

where the $X_i$ are the inputs and the partial derivatives are

$$\frac{\partial Q}{\partial X_i} > 0$$

$$\frac{\partial^2 Q}{\partial X_i^2} < 0.$$

With all input and output prices determined in competitive markets profits are simply total revenue minus total cost,

$$\pi = P_q Q - \sum_i P_i X_i.$$

Substituting for $Q$ we have

$$\pi = P_q f(X_1, \ldots, X_n) - \sum_i P_i X_i.$$

The first-order conditions for profit maximisation are

$$P_q f_i - P_i = 0 \qquad i = 1, 2, \ldots, n. \tag{8.1}$$

Rearranging this we have the usual condition for profit maximisation that the price of an input, $P_i$ (which is common to all producers), should equal the marginal revenue product resulting from a marginal change in the employment of that input, $P_q \partial Q / \partial X_i$.

For the case of net energy analysis we define $\alpha_i$ as the total energy input (both direct and indirect) used to produce one unit of input $i$. Net energy, $N$, is simply the difference between energy produced and energy consumed

$$N = Q - \sum_i \alpha_i X_i.$$

As before, substituting for $Q$

$$N = f(X_1, \ldots, X_n) - \sum_i \alpha_i X_i.$$

If the objective is the maximisation of net energy, then the first-order conditions are

$$f_i - \alpha_i = 0 \qquad i = 1, 2, \ldots, n \qquad (8.2)$$

or $\quad f_i = \alpha_i$,

which says that net energy will be maximised when marginal energy produced equals the marginal energy input.

Comparing the first-order conditions for the economic and net energy cases we have

$$P_i = f_i P_q \qquad \text{and} \qquad f_i = \alpha_i.$$

Substituting for $f_i$

$$P_i = \alpha_i P_q.$$

This says that economic analysis and net energy analysis would lead to the same allocations of resources if the prices of *all* inputs (including labour) were based on their energy content.

As stated by Hannon, this would be an energy theory of value.[26] The relative prices of all goods would be determined by their relative energy contents, that is

$$\frac{P_i}{P_j} = \frac{\alpha_i}{\alpha_j}$$

In this model energy is a homogeneous commodity and no distinction is made between energy as a stock resource or as a flow resource. The *only* scarce resource is assumed to be energy and relative prices would only change if the values of the $\alpha_i$s changed. Opportunity costs cease to be measured in terms of what a resource could earn in its next best paid use, instead they are measured in terms of energy requirements.

The previous analysis assumed a closed economy which was in equilibrium. It is now necessary to ask what would happen when the economy was in a disequilibrium situation, and to consider the effects of allowing for international trade. In the perfectly competitive model a change in demand and/

or supply conditions would lead to a change in the set of relative prices to reflect the changed relative scarcities of inputs and outputs. There will be time lags in the establishment of the new set of equilibrium prices caused by the time required to adjust capacity and production methods to the new market conditions. This means that in the short run quasi-rents may be earned and prices will deviate from relative energy contents. The market is signalling to consumers the changed relative scarcities of inputs and outputs. The market would be unable to perform this function if relative prices were based on relative energy contents. As Huettner has said: 'energy content pricing is an inefficient short-term pricing or resource allocation method'.[27] The market mechanism would have to be replaced by non-price-rationing systems, such as queueing and the use of points rationing systems.

This disequilibrium analysis clearly identifies the central problem with a pricing system under which relative prices are determined solely by reference to relative energy contents. This is that in the real world energy is not the only scarce resource, and that relative scarcities are changing over time. Once allowance is made for international trade this latter point assumes an even greater importance. Allowing for international trade causes no problems providing that there is universal perfect competition and all markets are in equilibrium. However, if some markets are in disequilibrium then international trade flows would be different if some countries established their relative prices via the market while others based them on energy inputs. In the international context, and allowing for problems of disequilibrium, it would probably be impossible for one country to go it alone and base its prices on relative energy contents.[28] If it did attempt this then almost certainly there would be a cost in terms of potential gross national product forgone. While G.N.P. has many deficiencies as a measure of welfare this potential loss of G.N.P. would still be pertinent information for public policy decisions on the appropriate basis for the setting of relative prices.

As previously mentioned, some energy analysts have suggested that, for example, both net energy analysis and economic analysis should be used in the assessment of decisions relating to the allocation of energy resources; for example, in the appraisal of energy research and development programmes, conservation programmes or proposals to develop new sources of fossil fuels. While it may be assumed that in many cases the same recommendation would be made, using both of these methods of analysis, it must also be assumed that in some cases they would conflict. Decision-making in those circumstances would require the allocation of weights to the alternative objective functions implied by the use of these different methods of analysis. In the absence of these weights there would be no consistent basis for the making of the required trade-offs.

## 8.5   Energy Analysis and Energy Conservation[29]

Although energy analysts in general recognise that energy analysis is a mechanistic technique without normative significance, some of them have suggested uses for this technique which imply that it has this significance. It has been suggested that one of the possible uses of energy analysis is the ranking of alternative-energy conservation projects.[30] These projects would be ranked in terms of the number of joules saved per £ invested.

Fundamental to the choice between, or ranking of, alternative policies is the specification in an operational form of an objective function. A prime concern of net energy studies is to ensure that in the development of some energy resource (e.g. shale oil or nuclear power) more energy is not invested than will be produced. This concern is sometimes expressed in statements relating to energy conservation. It is therefore worth considering whether energy conservation can be considered to be the (or an) objective of public policy.

Clearly when it is expressed in this general way the answer must be 'no' because it is non-operational. It does not specify whether all forms of energy are to be treated as equally scarce, or how much energy is to be conserved and over what time period and geographical area. Since all methods of production involve the use of energy should the economic

growth rate be chosen to maximise the rate of energy conservation? Since even a zero growth rate involves positive
production levels this would require that the chosen growth
rate be negative.[31] It seems more likely that the concern
with energy conservation is to be interpreted in terms of
maximising net energy, and therefore the selection of projects
so as to minimise the energy input of a given output.

There are a number of problems associated with this
approach, apart from the obvious one that it treats energy as
the only scarce resource. One is that this type of energy cost-
effectiveness analysis makes no allowance for social time-
preference. All energy conservation measures must have a
time-dimension, in that they will both take time to implement
and their effects will endure for a certain period. This immediately poses the problem of the determination of the length
of the planning period and the choice of the 'optimal' energy
(and capital) stock which is to be bequeathed to following
generations. This involves making value judgements.[32] Since
the effects of any policy will be uncertain it is necessary to
specify how risk and uncertainty are to be treated. In particular
it must be decided whether an error of, say, +100 GJ in the
estimate of energy requirements in any one year is to be
considered as being no worse than an error of −100 GJ with
the exception of the sign difference. This would be equivalent
to saying that the social marginal utilities of equal size gains
and losses were the same. But this seems unlikely to be the
case given the view of energy as the ultimate limiting resource.

We will now use a hypothetical example to illustrate some
of the problems involved in ranking alternative-conservation
measures on the basis of joules saved per £ invested. For
simplicity we assume that a choice is to be made between
three possible policies, all of which have zero operating costs,
the relevant data for which is given in Table 8.2.

Clearly policy *C* is inferior to policy *A* since in each year it
has higher-cost and lower-energy savings, and thus it can be
rejected. Policies *A* and *B* have the same aggregate investment
cost and the same aggregate energy savings. Their energy
savings per £ invested are identical. Does this mean that
society would be indifferent between them? The answer is
'no' unless society's marginal rate of time-preference is zero.

TABLE 8.2

*Alternative-Energy Conservation Policies*

| Policy/year | 1 £ | 2 £ | 3 £ | 4 GJ | 5 GJ | 6 GJ | 7 GJ | 8 GJ | 9 GJ | 10 GJ |
|---|---|---|---|---|---|---|---|---|---|---|
| *A* | 20 | 30 | 50 | 100 | 100 | 100 | 100 | 100 | 100 | 100 |
| *B* | 60 | 30 | 10 | 50 | 50 | 50 | 100 | 150 | 150 | 150 |
| *C* | 30 | 40 | 50 | 80 | 80 | 80 | 80 | 80 | 80 | 80 |

A positive rate of time-preference means that society prefers equal-size energy savings which are made relatively earlier to those which are made relatively later. Some of the problems involved in the choice of this rate were discussed in Chapter 3 (p. 67). For the present we shall arbitrarily assume that the time-preference rate is 10 per cent. Using this rate the discounted costs in year 1 of policies *A* and *B* are £88.57 and £95.53 respectively and their discounted energy savings are 402.3 GJ and 369.75 GJ respectively.[33] The ratio of discounted energy saved to discounted costs for policies *A* and *B* is 4.54 and 3.87 respectively and thus for the given data policy *A* is preferred. But what can we say about the suggested choice criterion and the data used in its implementation?

To return to a point we have made before, this criterion and data, with the use of the energy unit of account, implicitly assume that all forms of energy are equally scarce. That is, the choice criterion assumes that it is equally desirable to save 1 GJ of oil or 1 GJ of natural gas or 1 GJ of nuclear-generated electricity. The estimated reserves of these fuels, their different potentials for substitution given known technology and their different potential uses (use of oil as a fuel or as a feedstock into plastics) are all ignored.

In view of the alleged failings of market mechanisms which have been stressed by a number of energy analysts[34] it is important to note that all their strictures apply equally to this suggested choice criterion. The ranking of policies and projects using this criterion clearly depend on prices and other market conditions.

Allowing for the different relative scarcities of the various fuels it seems to us that the most useful input which energy analysis can provide to the appraisal of alternative-energy conservation policies would be the identification of the consequences for each separate fuel of the different policies. Although this procedure clearly encounters the adding-up problem it would have the advantage of providing decision-takers with useful and relevant information.

Clearly energy conservation can be furthered by switching from energy-intensive to non-energy-intensive products.[35] Energy analysis studies have, for example, shown that in the United States the bus is about one-third as energy intensive as the private car in intra-city operation and that total U.S. energy consumption could be reduced by about 5 per cent by a complete shift to the use of buses in cities.[36] As is typical with energy studies, products and services are assumed to have only one characteristic, in this case transportation between two places. Other characteristics, such as convenience, speed and comfort, are ignored. In addition, other inputs, such as travellers' time, are also ignored. However, the relevant point in the present context does not concern the characteristics of products, but is the basic question: why is energy analysis required to further a change from more to less energy intensive products? If consumer preferences are to count (which we have assumed in this book) then the price system can be used to signal to consumers and producers the changing relative prices of energy inputs. The use of the price mechanism in energy conservation programmes offers some indication of social preference for commodity switches, an indication which is not given by energy analysis. For an energy conservation programme information is required on how much energy costs would be saved by switching between products. This information could just as easily be obtained from a monetary input–output table as from the physical table of the energy analysis kind.

## 8.6    Energy Analysis and Investment Appraisal

The results of net energy analysis studies are sometimes presented in terms of the number of years which will elapse before the energy produced by a project would equal the energy consumed. For example, this has been done for both individual nuclear stations and for programmes of such stations in the United Kingdom.[37] An implication of these studies has been that the shorter is the period before net energy production occurs the better is the project. Economists will recognise the use of what is generally called the pay-back criterion.

This criterion has a number of deficiencies which must be understood before it is used in policy-making. These deficiencies apply equally whether the criterion is specified in value or in physical terms. In the latter case the criterion assumed that society is indifferent as to when energy consumption and production occur, and thus that a nominal unit of energy has the same worth irrespective of when its expenditure or saving occurs. Thus the criterion would rank the following two projects equally. (The negative signs indicate net energy consumption by the project and the positive signs net energy production.)

Each of these projects has a pay-back period of two years. However, even if society had a zero time-preference rate it would not be indifferent between them. With a zero or a positive rate of time-preference project *B* would be preferred to project *A*. The energy pay-back criterion ignores the timing of the net energy costs and benefits and in addition it ignores all the energy flows which occur after the pay-back date. This criterion is also unsatisfactory because it fails to normalise the alternative projects for their expected lives and capital outlays.[38] We therefore conclude that the information on the net energy flows which is generated by net energy studies should not be presented in terms of pay-back periods since the implicit criterion is unsatisfactory.

TABLE 8.3

Megajoules

| Project/year | 1 | 2 | 3 | 4 | 5 | 6 | 7 | 8 | 9 | 10 |
|---|---|---|---|---|---|---|---|---|---|---|
| A | −500 | −300 | −200 | 500 | 500 | 0 | 0 | 0 | 0 | 0 |
| B | 0 | 0 | −1000 | 500 | 500 | 500 | 500 | 500 | 500 | 500 |

## 8.7 Conclusions

In this chapter we have considered an analytical technique which some authors have suggested should, as a very minimum, be used to supplement economic analysis in the formulation of energy policies and in the making of choices concerning the allocation of energy resources. As we have seen, some authors have suggested that net energy analysis provides superior information to economic analysis for the making of these decisions. Since the U.S. Congress has mandated the use of net energy analysis we have argued that it is very important that its implications and limitations should be clearly understood.

We have argued that the use of net energy analysis as a practical tool for decision-taking is greatly reduced by the problems posed in defining the relevant sub-system and boundary. In addition we have shown that this technique suffers from many of the same problems as economic analysis; that its implicit objective function (maximisation of net energy) is very narrow and that it involves a view of the world in which for both the short- and long-term the only relevant resource constraint is that of energy. Finally we have argued that energy analysis and its variants lacks normative significance and thus should not be used to choose between alternative allocations of resources, whether they are presented as energy projects, conservation policies or in any other form.

# 9. ENERGY POLICY

## 9.1 Introduction

In this chapter we consider the formulation of energy policies in practice and bring together the analysis and discussion of previous chapters to assess the possible contribution of economics to the formation and implementation of a national energy policy. A feature of the 1970s has been the increasing concern of many governments and international institutions for the development of coherent energy policies. One of the principal aims of these policies has been the conservation of energy. This has a number of aspects, but basically it is concerned with the more efficient use of energy and a reduction in the amount of energy which is wasted. The stimulus for this concern would appear to have been partly the quadrupling of oil prices in late 1973 and early 1974 and the publication in the early 1970s of a number of computer-based simulation models of the future of the world.[1] The principal conclusion of these models was that the world 'system' will collapse in the twenty-first century due to overpopulation, resource depletion and pollution. Although the methodology and assumptions of these models have been soundly criticised by a number of economists[2] they received wide publicity and appeared to be taken seriously by some policy-makers in a number of countries.

The plan of this chapter is as follows. Some of the questions relating to the content and scope of an energy policy are considered in Section 9.2. Next, in Sections 9.3 and 9.4 respectively, recent developments in energy policy in the United Kingdom and in the United States are considered as two case studies. Finally the possible contribution of economics to the formulation and implementation of a national energy policy is reviewed and assessed in Section 9.5.

## 9.2 The Nature of Energy Policy

Energy policy is essentially concerned with the co-ordinated development of the separate fuel and power industries. From the economic point of view it is concerned with a number of interrelated questions, such as: How should the energy market be divided between the different available fuels, such as coal, oil, gas and electricity? What should be the relationship between the prices of the different fuels? At what rate should resources such as coal, natural gas and oil be depleted? How should related environmental factors be incorporated into energy policy decisions? What instruments should be used for the implementation of energy policy? These questions have been the subject of preceding chapters.

Energy policy must be formulated and implemented in a given and particular framework with numerous constraints. Policy in this sub-sector of the economy cannot be formulated in isolation from a nation's social and other objectives. Decisions taken within the energy industries will affect non-energy sector objectives. Thus the choice of pricing policy will affect the distribution of income; the planned balance between domestic and imported supplies will affect the balance of payments; the rate of run-down of a declining energy industry (such as coal in the United Kingdom in the 1960s) will affect both the aggregate level of unemployment and its regional distribution; the rate of increase of energy prices will affect the rate of inflation; and the revenue derived from the taxation of energy resources will, given a government's expenditure plans, have implications for other tax rates and affect the public sector borrowing requirement (with its implications for the growth of the money supply). In view of this interdependence between decisions taken in the energy industries and non-energy sector objectives, an important question concerns the extent to which these industries should be used to achieve these objectives.

Ideally the choice of an energy policy involves the selection of a course of action which is designed to achieve a particular set of objectives within an overall plan for the development of the energy sub-sector and for the economy as a whole.

The choice of policy should be accompanied by decisions on how it is to be implemented. This in turn requires the identification of the various decisions which have to be made, the determination of who is to make them, and thus the division of responsibility between the Government, the individual industries and consumers. In addition a choice has to be made on the appropriate organisational structures within which the decisions relating to the determination and implementation of energy policy are to be taken with a view to making it operational.[3]

There are many possible ways in which an energy policy could be operated. They vary from a complete reliance on the market mechanism with the implicit acceptance of the view that the allocation of resources should be determined by consumer preferences, to the other extreme where the state would determine the allocation through some administrative system which it substituted for the market. British energy policy has tended towards the former of these possibilities, but it has not relied on the unfettered workings of the market mechanism.


## 9.3 Energy Policy in the United Kingdom

Until the mid 1950s the British fuel economy was overwhelmingly dependent on coal, which supplied 90 per cent of the country's energy requirements in 1950, and because of this dependence on a single fuel Britain neither had nor required an energy policy at that time. The need for such a policy arose as the economy first shifted to a two-fuel basis with the growth in the importance of oil from the mid-1950s onwards, and to a four-fuel basis with the increased use of electricity from nuclear power stations and of natural gas during the 1960s.

Four official publications are of special importance to the development of U.K. energy policy in the 1960s and 1970s. These are the 1965 National Plan, the 1965 White Paper on Fuel Policy, the 1967 White Paper on Fuel Policy, and the 1978 Consultative Document on Energy Policy.

*The National Plan*

The National Plan, which was published in September 1965, was in part concerned with the development of an energy policy for the situation which faced the country in the mid-1960s.[4] The Plan recognised the importance of consumer choice in energy policy, but stated that attention must also be paid to other 'vital national interests', such as the balance of payments, security of supply and the health of the national fuel industries. The predominant concern of the latter was with the coal industry.

In 1954, because of doubts about the ability of the coal industry to meet demand, the Government had taken steps to encourage the use of oil, particularly in power stations.[5] But at the same time as these measures were being implemented there was a sudden fall in the demand for coal, and the electricity industry was asked to stop the substitution of oil for coal. Licences were refused for the import of American coal[6] and the 1961 Budget placed a duty of 2 old pence (~0.83 p) a gallon on oil used for burning. This tax was originally imposed for revenue reasons, but the main reason for its retention[7] was the protection which it gave to the coal industry, which in 1961 was approximately 40 per cent and equivalent to £1.15 per ton of coal used for steam-raising. These measures succeeded in halting the decline in the demand for coal between 1959 and 1963, although in those years most of the increase in the nation's fuel needs were met by oil, which increased its share in total energy consumption from 23 to 33 per cent.

The National Plan contained the first set of official forecasts for inland energy demand for a period five years ahead. Based upon the Plan's assumption that G.N.P. would increase by 25 per cent between 1964 and 1970 it was forecast that total inland energy consumption would be 324 million tons coal equivalent in 1970, which compares with a realised figure of 329 million tons. Although this forecast fortuitously (since the G.D.P. failed to grow at the forecast rate) turned out to be reasonably accurate, the forecasts overestimated the demand for coal by about 20 million tons.

## 1965 Fuel Policy

In 1965 the development of energy policy took a notable step forward with the publication of a White Paper on Fuel Policy[8] which set out to describe 'the principles which should govern a co-ordinated national fuel policy and the machinery and measures whereby the Government proposes to secure and maintain such a policy.'[9] An appraisal of energy policy was said to be timely, due to the technological advances which had occurred in the gas and electricity industries (with the low-cost production of gas from oil and the nuclear generation of electricity) and the problems facing the coal industry.

The White Paper's demand forecasts were taken from the National Plan. It stated that the single most difficult problem in energy policy was the health and size of the coal industry. It noted a continuing trend in relative energy prices in favour of oil and concluded that even if the decline in the demand for coal was to slow down in the 1970s it would be unlikely to be reversed. In view of this the White Paper announced some additional measures to assist the coal industry.[10] These were the provision by the Government of special funds to speed the closure of uneconomic pits by assisting the redeployment of labour in the industry and its resettlement in other industries; the write-off of £400 millions of the National Coal Board's £960 million debt to the exchequer; preference for coal over oil by the C.E.G.B. and the gas industry; and the promotion of the use of coal in public buildings. It is worth noting that the Government rejected the granting of a continuing subsidy to the coal industry, partly because, it was argued, this would reduce the incentive for the industry to be efficient and competitive.

With regard to oil the White Paper argued that there was little risk of oil supplies suddenly running out, although eventually oil prices could be expected to rise. The most serious problems associated with an increasing dependence on oil were seen as a potential risk to the security of supplies and increasing direct costs to the balance of payments. The Government stated that it proposed to lessen these potential risks by ensuring that the country held adequate stocks of oil, by encouraging the domestic refining of oil and by encouraging the oil companies to diversify their sources of supply (in effect risk-pooling).

Although the 1965 White Paper stated that it was concerned to set out the principles underlying the formation of energy policy, in fact it consisted basically of a review of the fuel industries while setting out some broad policy objectives.

## 1967 White Paper

The 1967 White Paper was in many ways a much more sophisticated document than its predecessor. The objective of energy policy was now stated as follows:

> The Government's aim is to see that our growing energy requirements are supplied in the way which yields the greatest benefit to the country. Policy for the fuel sector must therefore have regard to economic and social policy in other fields. In particular, the Government must ensure, through fuel policy, that national considerations which individual consumers do not take into account in choosing between competing fuels are given their due weight among the factors determining the pattern of fuel supply and demand. Such national considerations include security of supply, the efficient use of resources, the balance of payments, and the economic, social and human consequences of changes in the fuel supply pattern.[11]

It is to be noted that the objective now refers explicitly to the promotion of an efficient allocation of resources.

The reason for the 1967 reassessment of energy policy only two years after the publication of the previous White Paper was the fundamental change in the pattern of energy supply and demand resulting from the discovery of natural gas in the U.K. sector of the North Sea in the autumn of 1965 and the believed coming-of-age of commercial nuclear power. The stated aim of the new White Paper was to determine a set of long-term policy guidelines which might need to be modified if circumstances changed but which would provide a coherent framework for the taking of policy decisions.

The White Paper stressed that energy policy must be based on forecasts of what will happen on specified assumptions and with an appraisal of the alternative policy options. A notable feature of this White Paper was the considerable increase in the sophistication of the numerical techniques used by the Ministry of Power in the preparation of these forecasts.[12] The basic approach was to start with a series of statistical forecasts of the future pattern of energy demand given certain assumptions. Costs were then attached to each of these forecasts and the choice between them was based on the expected costs and the other factors specified in the fuel policy objective.

Two sets of estimates were presented in the White Paper, what were known as the 'January 1967 estimates' and the 'April 1967 estimates'. The first set of estimates was based on what was known as 'the assumptions exercise'.[13] The aim of this exercise was to produce unbiased estimates of the demand for fuel, using a set of assumptions which would embrace all the choices open to the Government. Variations were considered in two sets of assumptions. The first relating to the oil tax and the second to the quantity of natural gas which would be absorbed by certain dates. For example, with regard to the latter, it was assumed that either 2000 million cubic feet per day (mcfd) would be available in 1975 or 6000 mcfd.

The basic set of estimates was derived by calculating the total demand for all fuels together and then calculating the total for individual fuels by subdividing the total, using a sector by sector approach.

The 'January 1967 estimates' indicated that whatever policy was followed further substantial falls in the demand for coal were inevitable. It was forecast that the coal industry would face a number of very difficult problems even if the then current rate of protection as provided by the 2 old pence a gallon oil tax was left unchanged. By April 1967 additional information became available on the likely rate of absorption of natural gas. A new set of estimates was prepared (the 'April 1967 estimates') assuming the continuation of the existing oil tax and a 2000 mcfd supply of natural gas in 1970 and of 4000 mcfd in 1975. A summary of these

TABLE 9.1

*The April 1967 Estimates of Inland Demand for Energy*

|  | 1966 (actual) | 1970 | | 1975 | |
|---|---|---|---|---|---|
| Coal | 174.7 | 142* | (154.5) | 118 | (118.1) |
| Oil | 111.7 | 127 | (147.6) | 146 | (134.4) |
| Nuclear and hydro-electricity | 10.2 | 16 | (11.7) | 37 | (12.7) |
| Natural gas | 1.1 | 25 | (17.6) | 49 | (54.5) |
| Total | 297.7 | 310 | (331.4) | 350 | (319.7) |

*million tons coal equivalent (mtce)*

\* Assumed continued short-term support for coal.

SOURCES: Cmnd 3438, table D, and *Digest of U.K. Energy Statistics 1977*, Department of Energy (London: H.M.S.O., 1978) table 6.

estimates, with the actual out-turn figures shown in parentheses, is presented in Table 9.1.

The Government accepted the long-term trends indicated in Table 9.1 as a basis for planning. However, it considered that the forecast rapid decline in demand for coal between 1966 and 1970 would cause unmanageable problems for that industry, and it was decided to try to hold the demand for coal in 1970 to 155 million tons. Within the forecast total demand for energy of 310 mtce this was to be achieved by reducing the demand for oil to 125 mtce and for natural gas to 17 mtce. It was concluded that the only sector where it was practicable to plan for a significant increase in coal consumption was electricity generation. The Government proposed to reimburse the electricity industry for an extra coalburn of approximately 6 million tons a year. This constituted an important change in Government policy because previously the costs of additional use of coal by the electricity industry had fallen on electricity consumers.

The 'April 1967 estimates' had assumed that power stations would use 30 per cent of the available natural gas in 1970.

The decision to increase the total demand for coal in 1970 meant that it was likely that only three-quarters of the gas which it was assumed would be available would be absorbed in that year.

These estimates incorporated the results of certain decisions which had previously been made. The two most important of these concerned the Government's decisions on the planned depletion for natural gas from the North Sea and on the size of Britain's nuclear power programme.

The Government had decided to bring North Sea gas into use rapidly. Investment choice calculations based on a discount rate of 8 per cent had indicated that it would be more economic to convert existing appliances to use natural gas (with its higher heat content) than to convert it into town gas. Using a ten-year time-horizon it was estimated that the present value in 1968/9 of the capital costs of conversion would be £300 million, while the present value of the costs of the new plant which would be required if the system was not converted, together with the reforming costs of natural gas, would be over £600 million. Thus over a ten-year period conversion promised a present value saving of £300 million, and over thirty years the Gas Council estimated the saving to be £1400 million.[14]

The decision on which markets should be supplied with natural gas was taken by calculating where the resulting resource savings were likely to be greatest. Because of its characteristics of a clean, easy-to-control fuel which needed no storage by the consumer, manufactured gas had commanded a premium over the equivalent amount of heat from coal or oil in the domestic, commercial and certain industrial markets. As a result of this price premium these markets became referred to as 'premium markets'. Studies showed that resource savings would be highest if natural gas was sold in these premium markets and lowest in the bulk heat markets. Since, however, more gas was expected to be available than could be absorbed in the premium markets a rapid depletion rate required that some of it should be used in the bulk industrial and other non-premium markets.

Britain's first commercial nuclear power programme was announced in 1955, and after a number of changes it com-

prised nine Magnox stations with a capacity of 5000 MW.[15]
The second nuclear power programme was announced in
1965 and was based on the Advanced Gas-Cooled Reactor
(A.G.R.). Stations in this programme were estimated to have
generating costs lower than those of the best coal-fired
alternatives. This programme assumed that on average one
nuclear station would be commissioned each year from 1970
to 1975, giving a total of 8000 MW by 1975. Nuclear power
would then be providing 10 per cent of total energy requirements.

In the event the second nuclear power programme ran
into numerous difficulties and the first A.G.R. did not start
to produce electricity until 1976.[16] This accounts for the
substantially lower-than-forecast contribution of nuclear
energy to total energy requirements which is shown for 1975
in Table 9.1.

Before leaving this White Paper it is worth noting its view
on the supply of oil to the United Kingdom. It concluded
that there was sufficient oil in the ground to meet the world's
demands for well beyond 1980. The principal danger was
thought to be supply interruptions caused by political or
other events outside of the control of the Government or the
oil industry. However, it concluded that 'In the longer term
this danger is limited by the fact that producing countries
are at least as dependent on trade in oil as we are ourselves.'[17]
The White Paper noted that in the interest of security of
supply the United Kingdom has diversified its sources of
supply.

## 1978 Energy Policy Consultative Document

In the late 1970s the United Kingdom faced an energy
situation which was in many ways very different to that
considered in 1967 but which retained a number of the
problems which were considered at that time. The most
fundamental change concerned the discovery and exploitation
of commercial quantities of oil in the U.K. sector of the
North Sea. In February 1978 the Government set out its
latest proposals in a Green Paper.[18] Compared with the 1965
and 1967 White Papers the Consultative Document introduced

a number of important changes of both content and emphasis in energy policy.

First, recognising the long lead times associated with many energy projects, it adopted a longer planning horizon and provided detailed forecasts to the year 2000 and tentative ones into the twenty-first century. Second, learning from past errors, it rejected the adoption of a single energy policy and instead argued that the numerous uncertainties pointed to the need for a flexible strategy which could be kept continually under review and adapted to changed circumstances. Third, it placed more emphasis on the importance of global developments and the work of various international institutions to the formulation of energy policy in the United Kingdom. Fourth, although both the 1965 and 1967 White Papers had recognised the importance of the interdependence between the objectives of energy policy and other policy objectives of Government this interdependence was given greater emphasis in the 1978 Green Paper. Fifth, a wider range of policy instruments for the implementation of energy strategies was discussed.

Broadly speaking the objective of energy policy was still the provision of adequate[19] and secure supplies of energy at least social cost and the efficient allocation of resources. The function of energy policy was now stated to be government intervention to change the pattern of energy use, and hence the allocation of resources, which would result from relying solely on the working of the market mechanism in order to ensure that the energy economy develops in accordance with the national interest.[20] Energy-pricing policy was seen as one of the most important policy instruments for exerting this influence. This policy was to have regard to both the level and structure of prices, which should be related to long-run marginal costs. 'Energy prices should therefore at least cover the cost at which supplies can be provided on a continuing basis, while yielding an adequate return to investment.'[21] The Consultative Document stressed that consumers could be misled about the social costs imposed by their choice of fuels if all energy prices were not determined and set on the same basis.

The prime concern of the energy strategy set out in the

Consultative Document was to keep the options open. The need for taking a very long view of the future when planning energy strategy meant that the overriding concern was to take decisions which pointed the nation in the desired direction. The pervasive uncertainties surrounding energy policy meant that the approach of specifying a blueprint for the future (as in the 1967 White Paper) was rejected in favour of the adoption of a flexible strategy. As in 1967, however, this strategy had to be based on forecasts of energy supply and demand. Although the Green Paper emphasised the importance of measures designed to promote energy conservation it correctly concluded that even if these policies were successful and reduced the ratio of energy consumption to changes in G.D.P. total energy requirements would still increase along with increases in G.D.P. The need was thus to plan for increased energy requirements.

This planning was to take place in the context given by various predictions of a future severe physical shortage of world energy supplies (see Chapter 2, p. 10). The Government argued that, given these risks, it was worth paying substantial insurance premia to ensure adequate energy supplies in the future.[22] The size of such premia had necessarily to be limited and it was not possible to keep all the options open. The Government argued that in these circumstances the best course of action was to act in such a way that would be least costly over a wide range of possible outcomes, and would thus produce minimum regret.[23]

The proposed strategy which was set out in the Consultative Document repeatedly emphasised that the future was too uncertain for it to be possible to choose today a policy which would minimise the resource costs of meeting the forecast energy demands. It thus argued that those technological and manufacturing decisions should be taken which would keep a number of supply options open. This policy had implications for all four primary sources of energy in the United Kingdom.[24]

For the coal industry it meant the undertaking of an investment programme which would produce an extra 4 million tons a year from new and replacement capacity in the latter part of this century. Associated with these supply changes it was stated that action must be taken to ensure that markets exist for the coal, and in particular that the electricity-supply industry would continue to burn large quantities of coal.

For the generation of electricity from nuclear power stations it meant the maintenance of an adequate nuclear manufacturing industry which was capable of responding to a future demand for these stations. This capability was to encompass the construction of fast reactors since the Consultative Document concluded that on balance it was likely that they would be needed.

For oil it meant the choice of a depletion rate for supplies from the North Sea which would minimise the problems of switching to alternative supplies as they were run down.[25] In this connection the Government was concerned to avoid too sharp a peak and rate of depletion for both oil and natural gas.

The implementation of this strategy would involve a large investment in new energy sources from the late 1980s onwards to replace the declining output of oil and gas from the North Sea. The gradual adjustment of the energy economy away from oil was also seen to require action through international institutions, such as the International Energy Agency (I.E.A.) and the European Economic Community (E.E.C.).

The I.E.A. was formed in late 1974 as an autonomous body within the O.E.C.D.[26] Its main functions are the development of secure supplies of energy, the promotion of energy conservation and the operation of a scheme for sharing oil supplies in an emergency. The I.E.A. member countries have adopted the objective of limiting their oil imports to 26 million barrels a day in 1985. In support of this objective participating countries agreed in October 1977 to the adoption of a number of Principles. These included the formulation of national energy programmes which would encourage energy conservation and limit oil imports; the concentration of natural gas sales to premium markets; the adoption of pricing policies which would encourage energy conservation; and the expansion of nuclear generating capacity.[27]

The various treaties of the E.E.C. do not mention energy policy specifically. However, a Summit meeting held in Paris in October 1972 called for the early formulation of such a policy. The Council of Energy Ministers has met periodically since May 1973. In December 1974 the Council agreed to certain objectives for energy policy which the Community should seek to attain by 1985. A principal objective was the reduction of the Community's dependence on imported energy to a maximum of 50 per cent.

## 9.4　Energy Policy in the United States

In this section we limit our attention to the National Energy Plan which was presented to a joint session of Congress by President Carter on 20 April 1977.[28] This Plan constituted the most detailed attempt which has so far been made to formulate a comprehensive energy policy for the United States and at the same time provides some interesting examples of the ways in which economics might be used in the formulation and implementation of energy policy.

The background to the Energy Plan was the growing inbalance between the U.S. demands for and domestic supplies of oil and natural gas, and a consequent increasing reliance on imports of oil. Between 1950 and 1970 the real cost of energy in the United States fell by 28 per cent and this encouraged investment in a stock of capital goods, such as houses and cars which used energy inefficiently, and led to a rapid rate of increase in the consumption of energy. Energy consumption increased at an average annual rate of 3.5 per cent between 1950 and 1973. This growth rate was encouraged by the pricing policies of the Federal Government for natural gas and oil which held these prices below world price levels.[29]

In the 1970s the pricing policy for gas in inter-state commerce had to be accompanied by measures of physical rationing because of the increasing excess demand for gas at the regulated prices, and the available supplies of gas were not being concentrated on the premium markets. In 1976 the regulated well-head price of natural gas in inter-state commerce was only 25 per cent of the price of imported crude oil of the same heat value. Natural gas was in consequence a very attractive fuel for industry and electricity utilities. The

Federally regulated prices for natural gas in the inter-state markets were lower than those set in intra-state markets. The allocative significance of this was that the regulated prices discouraged the distribution of gas from the gas-producing states to other states, new gas production went primarily to the unregulated intra-state markets. Between 1973 and 1975 only 19 per cent of the additions to gas reserves were committed to inter-state markets.

The United States first became a net importer of oil in 1947. By 1974 it was importing 37 per cent of its oil consumption. The period between the quadrupling of oil prices in 1973/4 and the publication of the Energy Plan was characterised by an increasing dependence by the United States on oil imports. Imports of crude oil rose from 4.7 million barrels a day (mbd) in 1972 to 8.6 mbd in 1977, the latter figure representing over 40 per cent of consumption. During this period the domestic production of crude oil fell by roughly one-eighth. As with natural gas, Federal pricing policy for oil had been encouraging its consumption while discouraging exploration and production. In 1974 a policy of, in effect, taxing domestic oil production in order to subsidise oil imports was introduced.[30] Under this policy the Federal Energy Administration set an average price for domestically produced oil ($7.66 barrel in 1976). In order to refine this oil the producers had to purchase a ticket called an 'entitlement' at a cost of approximately $2 a barrel. This constituted a tax on domestic production. Imports were subsidised by granting refiners who imported oil at the (then) world price of about $12.5 a barrel an entitlement worth $3 a barrel. Whatever the source of the oil the cost to refiners was the same at $9.5 a barrel, and the price of oil was substantially below the world market price.[31]

The main objective of the Energy Plan was to reduce imports of crude oil and oil products and to limit the effects of interruptions to supply. The Plan, which made use of an econometric model, specified a number of objectives which were to be achieved by 1985. These included: a reduction in the annual growth of U.S. energy demand to less than 2 per cent; a reduction of oil imports in 1985 from a potential level of 16 mbd to 6 mbd; a 10 per cent reduction in the consumption of petroleum; the establishment of a strategic petroleum reserve of 1 billion barrels; the insulation of 90 per cent of all homes and other buildings; an increase in the annual production of coal by at least 400 million tons; and the establishment of a Department of Energy with responsibility for energy policy.[32]

The emphasis of the Energy Plan was on savings in energy consumption (and hence on energy conservation) and, except in the case of coal, it was only to a subsidiary extent concerned with increased production. Little encouragement was to be given to the increased production of oil and natural gas. The development of new energy sources and especially solar energy was, however, to be encouraged. A guiding principle of the Plan was that it should be equitable. Its proposals were also to take into account possible environmental effects. The implementation of the Plan was based on the extensive use of fiscal instruments (taxes and subsidies) and on the workings of the price mechanism.

The consumption of oil and natural gas was to be discouraged by the imposition of various taxes. Since gasoline consumption represents half of American total oil usage the Plan proposed the imposition of a graduated excise tax on what were called 'new gas-guzzlers'. It also proposed the introduction of a standby gasoline tax in the event of the gasoline consumption targets not being met. The tax would be 5 cents per gallon for each percent that consumption in the previous year exceeded the target. To encourage the substitution of coal for oil and natural gas by industrial users and the utilities a sliding-scale tax was proposed on their consumption of these fuels.

The Plan placed considerable emphasis on the role of prices in the implementation of energy policy and on the importance of the signalling function of the price mechanism. It noted that existing Federal pricing policies for natural gas and oil encouraged their consumption by keeping prices too low and tended to distort relative energy prices. To rectify this the Plan argued that the prices of natural gas and oil should reflect the fact that 'the true value of a depleting resource is the cost of replacing it'.[33] Having stated this principle the Plan then recoiled from its full implementation. The President argued that the total decontrol of oil and natural gas prices could have severe adverse effects on the U.S. economy and lead to large windfall profits for producers. To meet these potential difficulties the Plan proposed the phasing in over three years of a well-head tax on existing supplies of domestic oil to bring it to the level of the 1977 world price of oil. The price received by producers of previously discovered oil would remain at its existing level ($5.25 or $11.28 a barrel) except for adjustments for inflation. With regard to newly discovered oil its price would be allowed to rise over a three-year period to the 1977 world price, again with an allowance for inflation. Thus the Plan did not propose the full implementation of the pricing principle that the price of oil should be set with regard to its border price (unlike the governments of the United Kingdom and Australia which have accepted this principle – see Chapter 4, p. 100).

For the pricing of natural gas the Plan did appear, however, to accept the argument which was developed in Chapter 4, Section 4.8. The Plan proposed that beginning in 1978 the price limit for all new natural gas which was sold *anywhere* in the country should be set at the price of the equivalent energy value of domestic crude oil ($9.6 a barrel).[34] However, in terms of the argument of Chapter 4 it must be noted that this domestic price of oil would be less than the world price.

Finally on prices the Plan proposed a reform for utility rate structures for both gas and electricity. The Plan argued for the adoption of a common set of pricing principles by all gas and electricity utilities to avoid distortions in the signalling function of the price mechanism. It stated that 'Rates often do not reflect the costs imposed on society by the actions of utility consumers'. To rectify this the Plan proposed that both electricity and gas utilities should be required by law to phase out declining block tariffs which did not reflect costs. In essence the Plan argued for the setting of tariff levels and structures in relation to marginal costs. It wanted electricity utilities to be required to offer daily off-peak rates to all those consumers who were prepared to pay the associated increased metering costs, and for the utilities to offer tariffs with interruptible-supply clauses.[35]

It was argued in the Plan that the implementation of these pricing principles would permit the realisation of the Plan's equity objective. They would result in every consumer paying a fair share of the costs which his energy-consumption decisions placed on the nation.[36] The equity objective would also be pursued by, among other measures, reducing the unfairness of natural-gas pricing and by the dollar-for-dollar refund of the well-head tax on oil as it affected home-heating oil.

The Plan contained a number of proposals which were aimed at mitigating the possible adverse environmental effects of increased coal production and consumption. These included the required installation of the best available control technology in all new coal-fired plants and the eventual introduction of legislation for uniform national standards for strip mining.

The Energy Bill, which was approved by the 95th Congress in early October 1978, differed substantially from the proposals contained in the Energy Plan. Although Congress largely agreed with the objectives of the Energy Plan it refused to endorse many of President Carter's proposed measures to achieve them.[37] Notable omissions from the approved bill were the proposed tax on crude oil, the standby petrol tax, the tax on industrial and utility use of oil and natural gas, and the rebate of the gas-guzzling tax to buyers of fuel-economy cars.[38] The bill did approve the granting of tax credits for home insulation and the installation of solar- or wind-powered equipment, the reform of electricity rates for state public utilities – but this latter reform was not obligatory – and a tax on gas-guzzling cars beginning with 1980 models.[39] Under the terms of the Energy Bill the price of new natural gas will be decontrolled by 1985. The existing Federal price controls for natural gas are to be applied to both inter-state and intra-state supplies. On enactment of the bill the ceiling price of new gas was to rise from $1.50 per 1000 c.f. to $2.09, and subsequently to rise at an annual rate equal to the rate of inflation.

The most important omission from the Energy Bill was the proposed tax on crude oil. Even though the Energy Plan's proposals would not have resulted in the domestic price of crude oil being equal to the price of imports, the gap between these two prices would have been substantially reduced. If energy policy is to be based on the workings of the price mechanism then the proposed oil tax was probably necessary if the United States was to reduce its imports of oil sufficiently to allow the attainment of the I.E.A. import target for 1985. Almost certainly the measures which have been passed do not go far enough in reducing the United States' dependence on oil imports for the target to be realised.

## 9.5  Economics and Energy Policy

The previously considered case studies of energy policy in the United Kingdom and the United States have illustrated how economics has been used in practice in both the formulation and implementation of energy policy. It is perhaps worth pointing out that our concentration on economics should not be interpreted as implying that we believe that this is the only subject which has a contribution to make to the making of energy policy. Clearly this would be false. Since energy policy is concerned with the achievement of a number of objectives – economic, social, political, etc. – it follows that an interdisciplinary approach is required. As we noted in Chapter 6 (p. 144) it is partly the wide-ranging nature of many of these objectives which makes energy policy so complicated.

While recognising the need for an interdisciplinary approach it is our view that economics has a potentially important role to play in both the making and implementation of energy policies. At the same time, however, we are concerned that in this area it may be brought into disrepute by the formation of unrealistic and ill-informed expectations of what it can be expected to achieve. Nowhere is this more likely to happen than in the preparation and use of forecasts. Energy policies must be based on various forecasts of, for example, the demand for and supply of different fuels and of their prices. Clearly such forecasts must always be conditional and if they are to be useful for decision-taking they should not be point estimates but should consist of a range of estimates accompanied by the relevant set of confidence intervals. Unfortunately, published forecasts are often point estimates and their non-realisation is often (wrongly) interpreted as showing the limited use and value of economics.

Energy forecasts are often based on simple extrapolation and do not allow for the influence of prices on supply and demand. One reason for this appears to be that since late 1973 during the period of the so-called 'energy crisis' there has been much talk about energy being an essential commodity. But such talk usually confuses total and marginal utilities, as with the old 'paradox of value' in relation to water and diamonds.[40] At the margin energy is not essential but is subject to price elasticity, which, as we have seen in Chapter 4 (p. 78), is not equal to zero. The recognition of the importance of the distinction between total and marginal utility and of the responsiveness of both demand and supply to changes in price illustrates the potential use of elementary economic analysis in energy policy.

Another and important example of the potential value of elementary economic analysis relates to the Federal pricing scheme for natural gas which was introduced in 1960. As we have seen, in the 1970s the price of this gas was set considerably below the cost of oil of the equivalent heat value while gas prices in intra-state commerce were not subject to the Federal controls. In these circumstances, as we mentioned in Chapter 6 (p. 169), elementary demand and supply analysis could be used to predict the likely consequences of the chosen pricing policy. The theory would predict that, with a controlled price below the market clearing price, there would be excess demand, a need for physical (non-price) rationing of the available output, the discouragement of production and new exploration, and the concentration of available supplies in the intra-state markets.

This example illustrates one of the most valuable uses of economics in relation to energy policy. That is, the use of economic models to predict the likely consequences of particular policy decisions and a comparison of the relative efficacy of different policy instruments in terms of achieving particular objectives. In this connection we have explored, in

Chapter 6 (p. 146) the use of economics in the evaluation of different tax and subsidy instruments. As the analysis in that and other chapters has shown, another useful input of economics in this context is the identification of the informational requirements of different policies if they are to be made operational. A number of examples of this were provided in Chapters 3 and 4 (pp. 41, 105). In Chapter 3 we emphasised the importance of futures markets if inter-temporal resource allocation decisions were to be left solely to the workings of the market. In Chapter 4 we stated our view that while in principle energy prices could be used to achieve income distribution objectives the informational requirements of the optimising prices were so great as to make such prices non-operational.

The optimisation approach of economics to the development of energy policies – that is, the specification of a set of objectives and the use of economic analysis in association with a particular set of assumptions and definitions to identify the logical implications of those objectives – has been stressed in this book. In our view this approach is useful in both the tactical and strategic planning of energy policy. At the detailed planning level we have seen how in the context of U.K. energy policy economics was used as an aid to decision-taking with regard to the questions of which markets available supplies of natural gas should be supplied to and how oil supplies from the North Sea should be valued. In each case, assuming that the relevant objective was an efficient allocation of resources, it was seen how the concept of opportunity cost was used in the identification of the policy choice which would maximise expected resource savings. This was that gas should be supplied to the premium markets and that the oil should be valued at the border price.

This approach also has the advantage of clearly identifying when it is necessary to make value judgements, such as in relation to the distribution of income in the choice of pricing policies and the choice of the rate of discount.

Economics is useful in terms of many of the questions which face decision-makers in energy policy. It has an important role to play in clarifying issues and exploring the implications of the adoption of different policy options. One of its relative strengths in this as with other areas of policy is its stress on the word 'alternatives', not only in relation to alternative uses of resources but also in the potential application of different policy instruments to achieve given objectives.

# NOTES AND REFERENCES

## Chapter 2

1. The relative weights of different fuels in total energy production or consumption will vary with the system of aggregation which is used. Three measures of energy which are in common use are to measure it in terms of primary energy, on a heat-supplied basis, and as useful energy. Measurement in terms of primary fuels, e.g. coal, oil, natural gas, nuclear and hydro-electricity, involves the choice of a common unit of measurement such as coal or oil equivalent. With this system of measurement secondary fuels, such as town gas or electricity from coal-fired power stations, are measured in terms of the amounts of primary fuels required to make them. This means that secondary fuels are not then measured in terms of the quantities of such fuels actually consumed or purchased by the user sectors. This problem is overcome by measuring fuels on a heat-supplied basis; for example, in terms of Btu or therms. The heat content of secondary fuels is then shown as such, and the losses incurred in the conversion of primary to secondary fuels and in their distribution can be shown. However, this method does not deduct the additional losses which occur in the conversion into space heat, motive power, etc., by final users. The deduction of all these losses gives a measure of what is termed 'useful energy'. This method poses many measurement problems. For example, it requires reliable data on the purposes for which fuels are used in the different sectors and on the utilisation efficiencies of fuel-using appliances.

2. Some *approximate* energy equivalents are as follows:

7.3 barrels of crude oil = 1 tonne (= 2204.6 lb) of oil
1 million tonnes of oil = 39 million million Btu
                     = 395 million therms
                     = 10 thousand Teracalories
                     = 1.167 thousand million cubic metres of natural gas
                     = 41.2 thousand million cubic feet of natural gas
                     = 113 million cubic feet/day for a year natural gas
                     = 1.5 million tonnes coal equivalent

1 million barrels per day oil equivalent (mbdoe)
= 51 million tonnes oil equivalent per year
= 76 million tonnes coal equivalent per year.

The coal or oil equivalent of hydro or nuclear power stations is calculated in terms of the fuel input which would be required to produce the equivalent amount of electricity in fossil-fuelled power stations.

3. *Energy: Global Prospects 1985–2000*. Report of the Workshop on Alternative Energy Strategies (W.A.E.S.) (New York: McGraw-Hill, 1977) p. 169.

4. If present depletion policies are continued then gas supplies from indigenous gas reserves in Western Europe will decline before the end of this century. On the basis of present knowledge it appears unlikely that imports of gas will be available on the required scale at that time. Various studies have therefore argued that production from indigenous gas reservoirs should be limited. See, for example, B. de Vries and J. Kommandeur, 'Gas for Western Europe: How Much for How Long?', *Energy Policy*, vol. 3, no. 1 (Mar 1975) pp. 24–37.

5. 'A Generation of Change: World Energy Patterns 1950–1975', *Shell Briefing Service*, Apr 1976 (London: Shell International Petroleum Co., Shell Centre) table 2.

6. Ibid. p. 4.

7. See, for example, *Energy: Global Prospects 1985–2000*. Report of the Workshop on Alternative Energy Strategies (W.A.E.S.) (New York: McGraw-Hill, 1977); *Energy Policy – A Consultative Document*, Cmnd 7101 (London: H.M.S.O., 1978); *World Energy Outlook* (Paris: O.E.C.D., 1977); *International Energy Outlook 1985* (Washington: C.I.A., Library of Congress, 1977).

8. *International Energy Outlook 1985*, op. cit., and a report by the U.S. Department of Energy which suggested that a substantial gap between demand and supply might develop by 1985. See report in *The Times*, 22 Mar 1978, p. 21.

9. *Energy Policy: A Consultative Document*, op. cit. p. 9. More than half of the expected increase in capacity is located in Saudi Arabia.

10. *The National Energy Plan*, Executive Office of the President, The White House, 29 Apr 1977.

11. The following data show the percentage increases in fuel prices in the U.K. and the relative fall in the price of gas:

| Item | June 1966–Dec 1973 | Dec 1973–June 1976 |
|---|---|---|
| Electricity | 47 | 113 |
| Coal and coke | 82 | 78 |
| Gas | 29 | 47 |
| Retail price index | 61 | 59 |

12. For an analysis of the effects of Federal regulation of natural gas prices, see P. W. Macavoy and R. S. Pindyck, *The Economics of the Natural Gas Shortage (1960–1980)* (Amsterdam/New York: North-Holland/American Elsevier, 1975).

13. See E. W. Erickson and R. M. Spann, 'The US Petroleum Industry', in E. W. Erickson and L. Waverman (eds.), *The Energy Question*, vol. 2, *North America* (University of Toronto Press, 1974).

14. President Carter's *National Energy Plan* (Washington: Executive Office of the President, 29 Apr 1977) planned to reverse this trend by using various tax measures to raise the prices of gas and oil to industrial users and utilities in order to stimulate the demand for coal and other alternative fuels. The Plan also contained regulatory provisions to prohibit the burning of oil and natural gas in new utility and industrial boilers. See Chapter 9, Section 9.4.

15. See the discussion of this issue by C. Robinson, 'The Depletion of Energy Resources', in D. W. Pearce (ed.), *The Economics of Natural Resource Depletion* (London: Macmillan, 1975) p. 25.

16. The oil companies have been accused of being very conservative in their estimates of the reserves available in the North Sea. For an account and discussion of this accusation, see P. R. Odell, 'Optimal Development of the North Sea's Oil Fields – A Summary'; C. G. Wall *et al.*, 'Optimal Development of the North Sea's Oil Fields – The Criticisms'; P. R. Odell and E. Rosing, 'Optimal Development of the North Sea's Oil Fields – The Reply', all in *Energy Policy*, vol. 5, no. 4 (Dec 1977) pp. 282–306.

17. Robinson, op. cit., table 2.2.

18. This increase was mainly due to the use of what are known as secondary recovery techniques, which involve the pumping of water or gas into the oil reservoir to increase or maintain pressure. See *Energy: Global Prospects 1985–2000*. Report of the Workshop on Alternative Energy Strategies (New York: McGraw-Hill, 1977) p. 113.

19. See also the estimates given in the chapters on oil, coal, natural gas and nuclear energy in the W.A.E.S. report, op. cit.

20. The Institut Français de Petrole undertook a Delphi-type survey among leading world oil experts on future oil resources for the 1977 World Energy Conference, which was held in Istanbul. The outcome of this survey was an average estimate of 260 thousand million tonnes as the ultimately recoverable resources of conventional oil within the limit of a technical production cost gradually increasing to $20 – in 1976 dollars – per barrel in the year 2000. Ten per cent of the answers gave a pessimistic estimate of 175 thousand million tonnes, while 25 per cent of the experts were optimistic and gave answers in the range 350–475 thousand million tonnes. See the letter to *The Times* by Mr E. Ruttley, 29 Dec 1977.

21. See *Energy Policy: A Consultative Document*, op. cit.; *World Energy Outlook, O.E.C.D.*, op. cit.; *Energy Global Prospects 1985–2000*, op. cit.

22. *Oil and Gas Journal*, 29 Dec 1975. These reserve estimates are higher than those quoted in Table 2.9. Total world proven reserves at the end of 1975 were estimated in the *Oil and Gas Journal* to be about 90 thousand million tonnes.

23. 'World Energy Propsects', *Shell Briefing Service* (London: Shell

Centre, Oct 1977) p. 9. To put this figure in perspective the production cost, in the same value terms, of Arabian light crude oil in the Middle East was about 25 cents a barrel, while the production costs of most North Sea oil wells exceed $5 a barrel.

24. *Energy: Global Prospects 1985–2000*. Report of the Workshop on Alternative Energy Strategies (New York: McGraw-Hill: 1977) chap. 7.

25. 1 MW = 1000 kW.

26. Op. cit. p. 222.

27. *Energy: Global Prospects 1985–2000*, op. cit. p. 221.

28. Report from the U.S. Energy Research and Development Authority (ERDA), quoted in the W.A.E.S. Report, op. cit. p. 224.

29. 'World Energy Prospects', op. cit. p. 9

30. Op. cit. p. 225.

31. See 'The Exploitation of Tidal Power in the Severn Estuary', *Fourth Report from the Select Committee on Science and Technology*, Session 1976–7 (London: H.M.S.O. HCP 564, 1977).

32. One of the projects selected for large-scale tests is the Cockerell raft. This is a hinged platform which is designed to extract mechanical energy from the waves on which it floats and to convert it into electricity. It has been estimated that a single raft (measuring about 100 metres long and 50 metres wide) might generate 2 MW of electricity.

33. See 'The Prospects for the Generation of Electricity from Wind Energy in the United Kingdom', *Energy Paper Number 21* (London: H.M.S.O., 1977).

34. *Energy Policy: A Consultative Document*, Cmnd 7101 (London: H.M.S.O., 1977) p. 59.

## Chapter 3

1. See, for example, W. S. Jevons, *The Coal Question*, 3rd ed. (London: Macmillan, 1906). H. W. Richardson includes an interesting discussion of Jevons's work and the resource exhaustion controversy in general in *Economic Aspects of the Energy Crisis* (Farnborough, Hants: Lexington Books/Saxon House, 1975) chap. 2. On a more general level the subject of 'apocalyptic theory' is developed by Scott Gordon, 'Today's Apocalypses and Yesterday's', *American Economic Review* (May 1973) pp. 106–10.

2. The most famous example, of course, being D. H. Meadows *et al.*, *The Limits to Growth*. A Report for the Club of Rome's Project on the Predicament of Mankind (London: Pan Books, 1974).

3. See Chapter 2, Section 2.5.

4. E.g. N. Rosenberg, 'Innovative Responses to Materials Shortages', *American Economic Review* (May 1973) pp. 111–18. Also A. J. Surrey and William Page, 'Some Issues in the Current Debate about Energy and Natural Resources', in D. W. Pearce (ed.), *The Economics of Natural Resource Depletion* (London: Macmillan, 1975).

5. H. J. Barnett and C. Morse, *Scarcity and Growth* (Baltimore: Johns Hopkins University Press, 1963). Their findings are confirmed by W. Nordhaus and J. Tobin, 'Is Growth Obsolete?', *National Bureau of Economic Research*, Fiftieth Anniversary Colloqium (1972). See especially appx B on natural resources. They find that 'either the elasticity of substitution is high [between resources and a composite factor] or technological change is relatively resource-saving or both'. (p. 70).

6. Everything depends here on the view taken of technical progress. To some it appears rather as a sudden statistical 'event' random and unreliable. To others it is a predictable outcome of resource shortages and relative price changes. There is obviously some truth in both views, but where in the spectrum of opinion a writer belongs must inevitably colour his ideas on depletion rates.

7. A resource is 'essential' if without *some* of it as an input, output declines to zero This definition is used by P. Dasgupta and G. Heal, 'The Optimal Depletion of Exhaustible Resources', *Review of Economic Studies*, Symposium on the Economics of Exhaustible Resources (1974).

8. This should not be construed as saying that markets are notable for their inefficiency relative to other institutional arrangements. A feat of imagination is required to suppose that 'efficiency' could be achieved by any practical arrangements.

9. A. C. Pigou, *The Economics of Welfare* (London: Macmillan, 1924). F. P. Ramsey, 'A Mathematical Theory of Saving', *Economic Journal*, vol. 38 (1928) pp. 543–59.

10. The seminal work in this area is that of Harold Hotelling, 'The Economics of Exhaustible Resources', *Journal of Political Economy*, vol. 39 (Apr 1931) pp. 137–75. Earlier work had been done by L. C. Gray, 'Rent under the Assumption of Exhaustibility', *Quarterly Journal of Economics* (May 1914), reprinted as an appendix in M. Gaffney (ed.), *Extractive Resources and Taxation* (Madison: University of Wisconsin Press, 1967). More recent contributions include R. L. Gordon, 'A Reinterpretation of the Pure Theory of Exhaustion', *Journal of Political Economy* (June 1967), and A. T. Scott, 'The Theory of the Mine under Conditions of Certainty', in M. Gaffney (ed.), op. cit. pp. 25–62.

11. This term seems to have been coined by R. M. Solow in 'The Economics of Resources or the Resources of Economics', *American Economic Review*, Papers and Proceedings (May 1974).

12. Scott, 'The Theory of the Mine under Conditions of Certainty', op. cit. p. 34.

13. This diagrammatic analysis was first presented by O. C. Herfindahl, 'Depletion and Economic Theory', in Gaffney (ed.), *Extractive Resources and Taxation*, op. cit. pp. 63–90.

14. As price rises the output of the industry must clearly decline, and this had led to the assumption that all firms reduced their output accordingly. This phenomenon received the appellation 'tilting' since on

the assumption of U-shaped average-cost curves as in traditional theory each firm would no longer be producing at the lowest point. Schulze, however, has pointed out that, with capital mobility and perfect competition, the adjustment must come via changes in the number of firms in the industry. See W. D. Schulze, 'The Optimal Use of Non-renewable Resources: The Theory of Extraction', *Journal of Environmental Economics and Management* (1974) pp. 53–73.

15. R. L. Gordon, 'Conservation and the Theory of Exhaustible Resources', *Canadian Journal of Economics and Political Science* (1966) pp. 319–26.

16. Both, however, lead to a higher initial 'net price'.

17. For a more detailed analysis of these cases see Herfindahl, 'Depletion and Economic Theory', in Gaffney, op. cit.

18. For example as a result of 'pressure drop'.

19. This is obviously a gross oversimplification. After a certain point higher output rates may impose sharply higher costs. As pressure falls over time the marginal cost of output will rise and a faster rate of extraction may reduce the quantity of oil that can be finally extracted from a well.

20. See Herfindahl, 'Depletion and Economic Theory', in which the two-grade case is explored. Also Herfindahl and Kneese, *The Economic Theory of Natural Resources* (Columbus, Ohio: Charles E. Merrill, 1974) chap. 4.

21. Heal provides a model in which costs can rise with cumulative extraction up to a limit imposed by a 'backstop technology'. He concludes that 'instead of the price [of the resource] starting near marginal extraction costs and drawing exponentially away as royalty elements come to dominate, we find exactly the reverse'. See G. Heal, 'The Relationship between Price and Extraction Cost for a Resource with a Backstop Technology', *Bell Journal of Economics* (Autumn 1976) pp. 371–8.

22. The results stated here are most clearly derived in Schulze, 'The Optimal Use of Non-renewable Resources: The Theory of Extraction', *Journal of Environmental Economics and Management*, 1974. He concludes that 'there must exist an infinite number of future markets for the mineral resource over the continuum of grade, each market behaving similarly to the more simple case of a homogeneous resource with the value of a particular grade rising at the rate of interest' (p. 64).

23. A model of inter-generational Pareto efficiency is provided by T. Sandler and V. K. Smith, 'Intertemporal and Intergenerational Pareto Efficiency', *Journal of Environmental Economics and Management*, no. 2 (1976) pp. 151–9.

24. For a robust defence of the market from this point of view see J. Kay and J. Mirrlees, 'The Desirability of Natural Resource Depletion', in Pearce (ed.), *The Economics of Natural Resource Depletion*, op. cit., B.S.P. p. 163.

25. For example Schulze (1974) op. cit.

26. A more detailed proof is provided in T. Page, *Conservation and*

*Economic Efficiency: An Approach to Materials Policy* (Baltimore: Johns Hopkins University Press (1977) appx 9, p. 240.

27. Examples include Dasgupta and Heal, 'The Optimal Depletion of Exhaustible Resources', *Review of Economic Studies*, Symposium Issue 1974, op. cit., pp. 3–28; J. Stiglitz, 'Growth with Exhaustible Natural Resources: Efficient and Optimal Growth Paths' (same issue) pp. 123–37; and Heal, 'The Relationship between Price and Extraction Cost for a Resource with a Backstop Technology', *Bell Journal of Economics*, op. cit.

28. This sort of model is very common in the literature on economic growth. A corn economy is the most commonly used device to explain its features. Thus corn may be consumed now or used as a capital input in the production of next year's harvest.

29. See Ramsey (1928) op. cit.

30. E.g. Dasgupta and Heal (1974) op. cit., esp. pp. 13–18.

31. This is the definition of an 'essential resource' suggested by Dasgupta and Heal (1974) op. cit.

32. 'Bliss' was defined by Ramsey as a state in which the marginal utility of consumption or the marginal productivity of capital had declined to zero. Solow points out that Nordhaus's notion of a 'backstop technology' could be regarded as a dramatic way of asserting that $\sigma > 1$. *American Economic Review* (May 1974) p. 11, op. cit.

33. See, for example, A. C. Chiang, *Fundamental Methods of Mathematical Economics*, 2nd ed. (New York: McGraw-Hill, 1974) p. 419.

34. For the formal derivation of this result see Dasgupta and Heal (1974) op. cit. This example provides a particularly extreme demonstration of the effect of a positive pure time-preference rate. It appears that, if society discounts future utilities, it would optimally choose a consumption path declining toward zero eventually, even when production possibilities hold out the option of continually increasing consumption per head.

35. R. M. Solow and F. Y. Wan, 'Extraction Costs in the Theory of Exhaustible Resources', *Bell Journal of Economics* (Autumn 1976).

36. Schulze (1974) op. cit. pp. 58–64.

37. Dasgupta and Heal (1974) op. cit. pp. 18–25.

38. For an optimistic view see W. D. Nordhaus, 'Resources as a Constant on Growth', *American Economic Review*, Papers and Proceedings (May 1974) pp. 22–6.

39. J. Rawls, 'Justice as Fairness', in P. Laslett and W. G. Runciman (eds), *Philosophy, Politics and Society* (Oxford: Blackwell, 1962). Also *A Theory of Justice* (Oxford: Clarendon Press, 1972). Clearly it is impossible to summarise adequately Rawls's philosophy in a single paragraph.

40. Recent work in the field of optimal taxation closely mirrors the debate on inter-temporal equity, e.g. J. A. Mirrlees, 'An Exploration in the Field of Optimum Income Taxation', *Review of Economic Studies* (Apr 1971) pp. 175–208. Also A. B. Atkinson, 'How Progressive

282 *The Economics of Energy*

Should Income Tax Be', in M. Parkin (ed.), *Essays on Modern Economics* (London: Longman, 1973).

41. R. M. Solow, 'Intergenerational Equity and Exhaustible Resources', *Review of Economic Studies* (1974), Symposium Issue, pp. 29–45.

42. Solow recognises but avoids the question of where the initial stock of capital comes from.

43. This result is also derived by Stiglitz (1974) op. cit.

44. Ivor Pearce would seem to have some such criterion in mind, however, when he writes 'we might be thought to owe it to future generations not to take out what we cannot put back'. See 'Resource Conservation and the Market Mechanism', in Pearce (ed.), *The Economics of Natural Resource Depletion*, op. cit. p. 202.

45. Page, *Conservation and Economics Efficiency*, op. cit. chap. 8.

46. Herfindahl and Kneese, *Economic Theory of Natural Resources*, op. cit. p. 389.

47. For a useful survey see R. Lavard (ed.), *Cost Benefit Analysis* (Harmondsworth: Penguin Books, 1972) pp. 29–49.

48. Moral hazard is a problem which arises when states of the world are under the control, to some extent, of the person insured. Clearly it is more likely to pose serious difficulties in some spheres than in others. People are perhaps unlikely to kill themselves in order for their descendants to receive life-assurance benefits. They may, however, drive less carefully, work less hard, etc. For a good survey of this field see Herfindahl and Kneese (1974) op. cit. pp. 209–17. The major articles are K. J. Arrow and R. C. Lind, 'Uncertainty and the Evaluation of Public Investment Decisions', *American Economic Review* (June 1970) pp. 364–78. J. H. Hirshleifer and D. L. Shapiro, 'The Treatment of Risk and Uncertainty', *Quarterly Journal of Economics* (Feb 1963) pp. 95–111.

49. See Heal, 'Economic Aspects of Natural Resource Depletion', in Pearce (ed.), *Economics of Natural Resource Depletion*, op. cit. pp. 118–39.

50. A. K. Sen, 'On Optimizing the Rate of Saving', *Economic Journal* (Sep 1961) pp. 479–96.

51. S. A. Marglin, 'The Social Rate of Discount and the Optimal Rate of Investment', *Quarterly Journal of Economics* (Feb 1963) pp. 95–111.

52. See S. L. McDonald, *Petroleum Conservation in the U.S.: An Economic Analysis* (Baltimore: Johns Hopkins University Press, 1971).

53. We are assuming here that a company engaged in exploration has a selection of prospects with enough knowledge about each to make judgements about the probability of discovering commercially valuable deposits. See P. G. Bradley, 'Increasing Scarcity: The Case of Energy Resources', *American Economic Review* (May 1974) pp. 6–7.

54. See K. W. Dam, 'Oil and Gas Licensing and the North Sea', *Journal of Law and Economics* (Oct 1965), and 'The Evolution of North Sea Licensing Policy in Britain and Norway', *Journal of Law and Economics* (Oct 1974).

55. See Stiglitz, 'Growth with Exhaustible Natural Resources: The Competitive Economy', *Review of Economic Studies* Symposium Issue (1974) pp. 139—52.

56. Solow, 'The Economics of Resources or the Resources of Economics', *American Economic Review* (May 1974) pp. 6—7.

57. Kay and Mirrlees (1975) op. cit.

58. This proposition is proved formally in an appendix to Kay and Mirrlees (1975) op. cit.

59. For a critique of the 'nirvana' approach to economics see H. Demsetz, 'Information and Efficiency: Another Viewpoint', *Journal of Law and Economics* (Apr 1969) pp. 1—22.

## Chapter 4

1. See *Report of the Working Group on Energy Elasticities*, Energy Paper No. 17, Department of Energy (London: H.M.S.O., 1977) pp. 4—7.

2. Ibid. para. 9.

3. We are ignoring allocative changes caused by income effects. If the own price elasticity of demand equals zero an increase in price will still affect the allocation of resources in a general equilibrium model via associated income effects.

4. R. Halvorsen, 'Residential Demand for Electrical Energy', *Review of Economics and Statistics*, vol. 57, no. 1 (Feb 1975) pp. 12—18.

5. J. M. Griffin, 'The Effects of Higher Prices on Electricity Consumption', *Bell Journal of Economics*, vol. 5, no. 2 (Autumn 1974).

6. A. S. Deaton, 'The Measurement of Income and Price Elasticities', *European Economic Review*, vol. 6 (1975) pp. 261—73.

7. R. G. Hawkins, 'The Demand for Electricity: A Cross-Section Study of New South Wales and the Australian Capital Territory', *Economic Record* (Mar 1975).

8. See also W. D. Nordhaus, 'The Demand for Energy: An International Perspective', *Cowles Foundation Discussion Paper*, No. 405 (26 Sep 1975).

9. This principle is recognised in both U.K. and U.S. energy policy. See Chapter 9.

10. On the derivation and formulation of some of these rules see O. A. Davis and A. Whinston, 'Welfare Economics and the Theory of Second Best', *Review of Economic Studies* (Jan 1965); and O. A. Davis and A. Whinston, 'Piecemeal Policy in the Theory of Second Best', *Review of Economic Studies* (July 1967).

11. See, for example, R. Turvey, *Economic Analysis and Public Enterprises* (London: Allen & Unwin, 1971) chap. 3.

12. J. Hirshleifer, 'Peak Loads and Efficient Pricing: Comment', *Quarterly Journal of Economics* (Aug 1958) p. 458.

13. There is a large literature on the economics of peak-load pricing. Some useful references are: P. O. Steiner, 'Peak Loads and Efficient Pricing', *Quarterly Journal of Economics* (Nov 1957); O. E. Williamson,

'Peak Load Pricing and Optimal Capacity under Indivisibility Constraints', *American Economic Review* (Sep 1966); H. Mohring, 'The Peak Load Problem with Increasing Returns and Pricing Constraints', *American Economic Review* (Sep 1970); J. R. Nelson (ed.), *Marginal Cost Pricing in Practice* (Englewood Cliffs, N.J.: Prentice-Hall, 1964) (this volume contains translations of many of the most important papers by French writers on the theory underlying the Green Tariff of Electricité de France); I. Pressman, 'A Mathematical Formulation of the Peak Load Problem', *Bell Journal of Economics and Management Science*, vol. 1, no. 2 (Autumn 1970); M. A. Crew and P. E. Kleindorfer, 'Peak Load Pricing with a Diverse Technology', *Bell Journal of Economics*, vol. 7, no. 1 (Spring 1976).

14. For an analysis of the effects of allowing for interdependent demand functions see Steiner, op. cit. pp. 608–9, and Pressman, op. cit.

15. See R. Turvey, *Optimal Pricing and Investment in Electricity Supply* (London: Allen & Unwin, 1968); and J. T. Wenders, 'Peak Load Pricing in the Electric Utility Industry', *Bell Journal of Economics*, vol. 7, no. 1 (Spring 1976).

16. For an analysis of the effects of allowing for uncertainty in the determination of the optimal peak and off-peak prices see Crew and Kleindorfer, op. cit.; also G. Brown and M. B. Johnson, 'Public Utility Pricing and Output under Risk', *American Economic Review* (Mar 1969), and G. Brown and M. B. Johnson, 'Public Utility Pricing and Output under Risk: Reply', *American Economic Review* (June 1970). See also J. C. Panzar and D. S. Sibley, 'Public Utility Pricing under Risk', *American Economic Review* (Dec 1978).

17. See H. S. E. Gravelle, 'The Peak Load Problem with Feasible Storage', *Economic Journal* (June 1976); D. T. Nguyen, 'The Problem of Peak Loads and Inventories', *Bell Journal of Economics*, vol. 7, no. 1 (Spring 1976); and G. Pyatt, 'Marginal Costs, Prices and Storage', *Economic Journal* (Dec 1978).

18. See R. Turvey, 'Marginal Cost Pricing in Practice', *Economica* (Nov 1964); and M. G. Webb, *The Economics of Nationalised Industries* (London: Nelson, 1973) pp. 100–1 and 114–15.

19. See Gravelle, op. cit. p. 265.

20. The determination of whether $0\overline{q}$ is the optimal capacity in the presence of feasible storable is complicated. See Gravelle, op. cit. pp. 267–8.

21. In this connection it is interesting to note that the economists at Electricité de France have claimed that they can calculate the relevant marginal costs simply by reference to marginal operating costs on the assumption that the investment programme is optimal: 'when the capacity of fixed assets assures least-cost production, the cost of expansion is equal to the marginal cost of operation of existing plants'. See M. Boiteux and P. Stasi, 'The Determination of Costs of Expansion of an Interconnected System of Production and Distribution of Electricity', in J. R. Nelson (ed.), *Marginal Cost Pricing in Practice* (Englewood Cliffs, N.J.: Prentice-Hall International, 1964) chap. 5, p. 94.

22. On this issue see R. Turvey and D. Anderson, *Electricity Econ-*

*omics* (Baltimore and London: Johns Hopkins University Press, 1977). A rather less detailed and simpler discussion of the design of electricity tariffs will be found in M. G. Webb, *Power Sector Planning Manual* (London: Ministry of Overseas Development, 1979).

23. This is the case with the White Meter Tariff in England and Wales. In April 1977 the terms of this tariff for the North Eastern Electricity Board were that there was a quarterly charge of £3.74, plus a charge of 2.61 p. for each unit taken between the hours of 7 a.m. and 11 p.m. of each day and of 1.06 p. for each unit taken at other times. These charges were subject to a fuel adjustment clause to reflect increases in the Central Electricity Generating Board's fuel costs.

24. See Webb, *Power Sector Planning Manual*, op. cit.

25. On the derivation of these weights see Turvey, *Economic Analysis and Public Enterprises*, op. cit. p. 32.

26. See 'Domestic Tariffs Experiment', *Load and Market Research Report*, No. 21 (London: Electricity Council, 1975); and J. G. Boggis, 'An Electricity Pricing Experiment in England and Wales', in H. M. Trebing (ed.), *New Dimensions in Public Utility Pricing* (East Lansing, Mich.: Michigan State University Public Utilities Studies, 1976).

27. *The National Energy Plan* (Washington, D.C.: Executive Office of the President, The White House, 29 Apr 1977) p. 46.

28. *Gas and Electricity Prices, Fourth Report From the Select Committee on Nationalised Industries*, Session 1975–76 (London: H.M.S.O., 1976), House of Commons Paper 353, p. 322.

29. One Megawatt (MW) = 100 kilowatts (kW).

30. R. A. Peddie, 'Peak Lopping and Load Shaping of the C.E.G.B.'s Demand', *Seventeenth Hunter Memorial Lecture,* 9 Jan 1975, (London: Institution of Electrical Engineers).

31. Y. Balasko, 'A Contribution to the History of the Green Tariff: Its Impact and Its Prospects', in Trebing (ed.), *New Dimensions in Public Utility Pricing*, op. cit.

32. This view has been put forward by a number of authors. See, for example, M. G. Webb, *Pricing Policies for Public Enterprises* (London: Macmillian, 1976); and W. Vickrey in Trebing (ed.), *New Dimensions in Public Utility Pricing*, op. cit. p. 547.

33. A good example of the application of system planning techniques and their relationship to optimal prices will be found in *Gas Prices (Second Report)*, Report No. 102 of the National Board for Prices and Incomes, Cmnd 3924 (London: H.M.S.O., 1969).

34. See, for example, Turvey, *Optimal Pricing and Investment in Electricity Supply*, op. cit.; and R. W. Bates and N. Fraser, *Investment Decisions in the Nationalised Fuel Industries* (London: Cambridge University Press, 1974) chap. 5.

35. For discussions of this problem see Bates and Fraser, op. cit. chaps 3 and 4; Turvey and Anderson, op. cit. chap. 13; and Webb, *The Economics of Nationalised Industries*, op. cit. chap. 5.

36. See Turvey, 'Marginal Cost', *Economic Journal* (June 1969).

37. See, for example, Turvey and Anderson, op. cit. chap. 5.

38. For an interesting exercise in the calculation of these shadow

prices for the U.K. energy sector before the 1973 oil price rises see Michael Posner, *Fuel Policy: A Study in Applied Economics* (London: Macmillan, 1973).

39. I. M. D. Little and J. Mirrlees, *Project Appraisal and Planning for Developing Countries* (London: Heinemann, 1974); and also Lyn Squire and H. G. van der Tak, *Economic Analysis of Projects*, A World Bank Research Publication (Baltimore and London: Johns Hopkins University Press, 1975).

40. Oil from the U.K. continental shelf is valued at the price of imported light crude oil. The Australian budget for 1978/79 introduced a measure to price all Australian-produced crude oil at refineries at the same price as imports.

41. It is interesting to compare the conclusions of this analysis with the pricing proposals for domestic oil and gas which were contained in *The National Energy Plan* for the United States, op. cit. pp. 50–5. The Plan stated that 'There is little or no basis for the assertion that the only reasonable price for all domestic production is the world oil price'. However, via a combination of measures relating to the well-head price of oil and the taxation of oil consumers would have been faced with a price of oil which reflected the border price in 1977. See Chapter 9.

42. On the economics of regulation of private utilities see E. E. Bailey, *Economic Theory of Regulatory Constraint* (Lexington, Mass.: Lexington Books, D. C. Heath, 1973).

43. See W. J. Baumol and D. F. Bradford, 'Optimal Departures from Marginal Cost Pricing', *American Economic Review* (June 1970) vol. 60, no. 3.

44. See R. Rees, *Public Enterprise Economics* (London: Weidenfeld & Nicolson, 1976) p. 105.

45. Notice that at a practical level this analysis suggests that if an electricity or gas utility was selling its output using two-part tariffs then it would be preferable to change the standing charge rather than the unit rate. The extra revenue would then be paid out of the consumers' surpluses.

46. See Rees, op. cit. p. 107.

47. See, for example, M. S. Feldstein, 'Distributional Equity and the Optimal Structure of Public Prices', *American Economic Review*, vol. 62, no. 1 (Mar 1972) pp. 32–6, and M. S. Feldstein, 'Equity and Efficiency in Public Sector Pricing: The Optimal Two-Part Tariff', *Quarterly Journal of Economics*, vol. 86, no. 2 (May 1972) pp. 175–87.

48. It has been estimated that in Great Britain in the year April 1975 to March 1976 poorer households with average incomes of £1300 had fuel costs which represented 10% of their income, while for richer households with average incomes of £4500 fuel costs took only 5% of income. See Julia Field and B. Hedges, *National Fuel and Heating Survey* (London: Social and Community Planning Research, 1977).

49. A free-allowance scheme for electricity was introduced in Ireland in 1967. Basically the scheme applies to old-age pensioners who are in receipt of social security benefits and who are registered consumers of electricity. Under the scheme the normal two-monthly fixed

charge is waived together with the cost of up to 200 to 300 units of electricity per two-monthly billing period in the summer and winter periods respectively. By December 1975 12% of all domestic consumers of electricity were receiving the allowance at a total cost in the 1975 calendar year of just over £2.5 million. The use of a free-allowance scheme for the United Kingdom was rejected by an inter-departmental group of officials, who produced a report *Energy Tariffs and the Poor* (London: Department of Energy, 1976).

50. Inverted tariffs for electricity were introduced in Japan in 1974 and in California in 1975.

51. See, for example, J. W. Howe, 'Lifeline Rates – Benefits for Whom?', *Public Utilities Fortnightly*, vol. 97, no. 3 (1976) pp. 22–5; and *Paying for Fuel*, National Consumer Council Report No. 2 (London: H.M.S.O., 1976).

52. J. W. Wilson and R. G. Ukler, 'Inverted Electric Utilities Rate Structures: An Empirical Analysis', in Trebing (ed.), *New Dimensions in Public Utility Pricing*, op. cit.

53. Unless the social welfare payments were financed through the imposition of lump-sum taxes (which is not feasible) the taxes raised to pay for them would introduce additional distortions into the price system. If income taxes were raised there could be adverse effects on the supply of work effort and on the willingness to take risks. Thus ideally the choice between alternative methods of achieving the desired distributional objectives should be in terms of a general equilibrium analysis and of a comparison of the welfare losses resulting from each of the possible redistributional policies. That policy option which achieved the desired result with the minimum loss of welfare (both direct in the industry concerned and indirect through the raising of taxes) would be preferred.

54. In 1976 the U.K. Government announced a scheme whereby consumers of electricity who were receiving either supplementary benefits or family income supplement and who paid directly for their electricity (either via quarterly bills or slot meters) could claim a quarter off their bills during the February to April meter-reading quarter. Although over 2½ million people were eligible by 18 Mar 1977 only 238,000 had claimed their discounts. The average discount was between £6 and £7.

55. In some countries these policy options may not be available because of the absence of a well-developed social security system. This is the case in many developing countries. The manipulation of energy prices may then be the best available policy instrument to achieve distributional objectives.

# Chapter 5

1. N. C. Rasmussen, 'How Safe is Nuclear Power?', talk given at the Thomas Alva Edison Foundation in the United States and reprinted in

*Atom*, no. 252 (Oct 1977), information bulletin of the U.K. Atomic Energy Authority, London.

2. Reported in *The Guardian*, 26 Nov 1977, p. 4.

3. L. B. Lave and E. Seskin, 'Air Pollution and Human Health', *Science*, vol. 169 (21 Aug 1970) pp. 723–31.

4. R. Wilson and W. J. Jones, *Energy, Ecology and the Environment* (New York, San Francisco and London: Academic Press, 1974) p. 188.

5. *National Coal Board Report on Accounts, 1976/7* (London: H.M.S.O., p. 15).

6. A. D. Bradshaw and M. J. Chadwick, *The Restoration of Land* (Oxford: Blackwell 1978) tables 2.5 and 2.6.

7. Garvey, *Energy, Ecology and Economy* (New York: Norton, 1972) p. 97.

8. A study prepared by the Inter-Governmental Maritime Consultative Organisation (IMCO) in 1978 showed that in the period 1968–77 there were over 200 serious tanker fires or explosions, killing over 450 people. See *The Economist*, 13 Jan 1979, p. 67.

9. C. Cicchetti and A. M. Freeman III, 'The Trans-Alaska Pipeline: An Economic Analysis of Alternatives', in A. C. Enthoven and A. M. Freeman III (eds), *Pollution, Resources, and the Environment* (New York: Norton, 1973).

10. Energy Commission Paper no. 1, *Working Document on Energy Policy* (London: The Energy Commission, 1977) p. 65.

11. Royal Commission on Environmental Pollution, Sixth Report, *Nuclear Power and the Environment*, Cmnd 6618 (London: H.M.S.O., 1976).

12. Ibid. p. 16.

13. Ibid. p. 19.

14. U.S. Atomic Energy Commission, *Reactor Safety Study*, WASH–1400 (Washington, D.C.: Government Printing Office, 1 74).

15. Rasmussen, 'How Safe is Nuclear Power?', talk given at the Thomas Alva Edison Foundation in the United States and reprinted in *Atom*, no. 252 (Oct 1977), information bulletin of the U.K. Atomic Energy Authority, London.

16. U.K. Atomic Energy Authority, *Fourth Annual Report 1957–58* (London: H.M.S.O., 1958) p. 9.

17. Wilson and Jones, *Energy, Ecology and the Environment*, op. cit. p. 260.

18. For example, on 9 February 1978 a dam collapsed in the Big Tujunga Canyon near Los Angeles, resulting in nine deaths.

19. Wilson and Jones, op. cit. p. 262.

20. M. A. McKean, 'Japan', in D. R. Kelley (ed.), *The Energy Crisis and the Environment* (New York and London: Praeger, 1977) p. 173.

21. Royal Commission on Environmental Pollution, *First Report*, Cmnd 4585 (London: H.M.S.O., 1971).

22. See P. Chapman, *Fuel's Paradise: Energy Options for Britain* (Harmondsworth: Penguin Books, 1975) chap. 6.

23. M. R. Dohan and P. F. Palmedo, 'The Effect of Specific Energy

Uses on Air Pollutant Emissions in New York City 1970–1985', BNL–19064 (Brookhaven National Laboratory, Upton, N.J.: January 1974).

24. *1974 Profile of Air Pollution Control* (Air Pollution Control District, Country of Los Angeles) p. 69.

25. W. T. Mikolowksy *et al.*, *The Regional Impacts of Near Term Transportation Alternatives: A Case Study of Los Angeles* (Santa Monica: The Rand Corporation, Report no. R–1524 – S.C.A.G., June 1974).

26. *Plan to Protect the Inhabitants of Tokyo Metropolis from Public Hazard* (Tokyo, 1974), quoted in *Environmental and Energy Use in Urban Areas* (Paris: O.E.C.D., 1978) p. 35.

27. R. Coase, 'The Problem of Social Cost', *Journal of Law and Economics*, vol. 3 (1960) pp. 1–44.

28. *The Polluter Pays Principle: Definition, Analysis, Implementation* (Paris: Organisation for Economic Co-operation and Development, 1975).

29. T. H. Tietenberg, 'On Taxation and the Control of Externalities: Comment', *American Economic Review* (June 1974), and Royal Commission on Environmental Pollution, Fifth Report, *Air Pollution Control: An Intergrated Approach*, Cmnd 6371 (London: H.M.S.O., 1976) paras 35–40 and 337–41.

30. Note that the optimal tax should be applied to the emission and not to the unit of output. The aim of the tax is not to affect a firm's output but its choice of inputs. See C. L. Plott, 'Externalities and Corrective Taxes', *Economica* (Feb 1966) pp. 84–7.

31. For a discussion of the informational requirements of standards versus taxes see P. Burrows, 'Pricing versus Regulation for Environmental Protection', in A. J. Culyer (ed.), *Economic Policies and Social Goals* (London: Martin Robertson, 1974).

32. See *Economic Measure of Environmental Damage* (Paris: Organisation for Economic Co-operation and Development, 1976), and *Pollution Charges: An Assessment* (Paris: O.E.C.D., 1976).

33. W. J. Baumol and W. E. Oates, 'The Use of Standards and Prices for the Protection of the Environment', *Swedish Journal of Economics* (Mar 1971).

34. See Baumol and Oates, op. cit., and W. J. Baumol, 'On Taxation and the Control of Externalities', *American Economic Review* (June 1972).

35. See Tietenberg, op. cit.

36. For an analysis of the problem of two target variables in pollution control see M. Ricketts and M. G. Webb, 'Pricing and Standards in the Control of Pollution', *Scandinavian Journal of Economics*, vol. 80, no. 1 (1978).

37. *Pollution Charges: An Assessment* (Paris: O.E.C.D., 1976) p. 37.

38. Royal Commission on Environmental Pollution, Fifth Report, *Air Pollution Control: An Integrated Approach*, Cmnd 6371 (London: H.M.S.O., 1976). para. 53.

39. In this case the minimisation of transactions costs and the

requirement for ease of understanding argue strongly for the pollution tax to be levied on domestic quality coal rather than on smoke emissions.

40. The householder decides which form of fuel he will use to conform to the order (e.g. smokeless fuel, gas, electricity). Usually 70% of the cost of conversion is refunded to the householder by the local authority.

41. D. M. Fort *et al.*, 'Proposal for a Smog Tax', reprinted in *Tax Recommendations of the President*, Hearings before the House Committee on Ways and Means, 91st Congress, 2nd Session, 1970, pp. 369—79. The account in the text is taken from A. V. Kneese, *Economics and the Environment* (Harmondsworth: Penguin Books, 1977) p. 215.

42. For a classification and description of the different types of environmental damage see K. G. Maler and R. E. Wyzga, *Economic Measurement of Environmental Damage* (Paris: O.E.C.D., 1976) chap. 4.

43. For a useful discussion and summary of a number of studies of the effects of air pollutants on man see *Air Quality Criteria and Guides for Urban Air Pollutants*; Report of a W.H.O. Expert Committee, World Health Organisation Technical Report no. 506 (Geneva, 1972).

44. J. R. Hicks, 'The Four Consumers' Surpluses', *Review of Economic Studies*, no. 1 (1943).

45. See J. R. Hicks, *A Revision of Demand Theory* (Oxford: Clarendon Press, 1956) chap. 8.

46. For a very useful discussion of these issues see D. W. Pearce, *Valuing the Damage Costs of Environmental Pollution*, Report to the U.K. Department of the Environment (London: Department of the Environment, 1976).

47. Note that there may be a problem of the non-separability of the various emissions, as with smog.

48. See, for example, R. W. Johnson and G. M. Brown Jr, *Cleaning up Europe's Waters* (New York: Praeger, 1976).

49. *Pollution Charges: An Assessment*, op. cit. pp. 41—3.

50. Ibid. pp. 43—5.

51. *S.C.A.G. Short Range Transportation Plan* (Los Angeles: Southern California Association of Governments, 11 Apr 1974).

52. In this connection note that the analysis of Section 5.4 would need to be amended to allow for monopolies. If the price elasticity of demand approximates to zero then the industry could simply pass a pollution tax forward to consumers and thus leave the level of emissions unchanged. In these circumstances direct controls may be preferable to the use of the tax instrument.

## Chapter 6

1. See, for example, L. Johanson, *Public Economics* (New York: North-Holland, 1965) pp. 9—22, and A. Peacock and G. K. Shaw, *The*

*Economic Theory of Fiscal Policy* (London: Allen & Unwin, 1971) chap. viii.

2. This may be required where oil or gas fields straddle national boundaries. In the North Sea, for example, the discovery of the Brent field gave rise to classic externality problems. In spite of the 'go-slow' policy of Norway towards oil development, production licences were awarded since the reservoir 'was especially conducive to draining of the Norwegian Oil by pumping from the British side'. See K. W. Dam, 'The Evolution of North Sea Licensing Policy in Britain and Norway', *Journal of Law and Economics*, vol. 17. no. 2 (Oct 1974) p. 234.

3. Unfortunately the term 'royalty' can have different meanings. In Chapter 3 we used the word 'royalty' as referring to the capitalised value of the rents derivable from working a deposit – and hence to the value of those deposits in the market. In this chapter we shall also have occasion to use the more usual definition of the term as a levy on the value of output paid to the licensor by the licensee whether the former is a government or a private individual.

4. In practice, of course, some assets would tend to be more favourably treated than others. The obvious example is 'human capital' which would be very difficult to tax. Further, this statement about neutrality should not be construed as denying the possibility of effects via changes in savings propensities.

5. The basic issue at stake is whether the corporation tax falls on profits or whether it is shifted forward to consumers in the form of high prices. If companies are profit-maximisers and the tax falls on pure profits one argument runs that no shifting will take place since whatever price and output combination maximised profits before the tax was introduced would also do so afterwards. An alternative view concentrates on behavioural theories of the firm and the existence of mark-up pricing, along with the fact that profits taxes may fall on 'cost' items, to make the case for forward shifting. Much empirical work has failed to settle the issue one way or the other. For a survey of the evidence see R. A. Musgrave and P. B. Musgrave, *Public Finance in Theory and Practice*, 2nd ed. (New York: McGraw-Hill, 1976) chap. 18.

6. See S. L. McDonald, 'Percentage Depletion, Expensing of Intangibles, and Petroleum Conservation', in M. Gaffney (ed.), *Extractive Resources and Taxation* (Madison: University of Wisconsin Press, 1967) pp. 269–88.

7. This point is emphasised by J. E. Stiglitz, 'The Efficiency of Market Prices in Long-run Allocations in the Oil Industry', in G. M. Brannon (ed.), *Studies in Energy Tax Policy* (New York: Ballinger, 1975), pp. 77–8.

8. This is a very important and complex problem for any system of income taxation to solve. See R. A. Musgrave, *The Theory of Public Finance* (New York: McGraw-Hill, 1959) chap. 8, for a survey of the problem involved in defining 'taxable income'. The present U.S. corporation income tax is described in J. A. Pechman, *Federal Tax Policy*, 3rd ed. (Washington, D.C.: Brookings Institution, 1977) chap.

5. For the U.K. system see A. R. Prest, *Public Finance in Theory and Practice*, 5th ed. (London: Weidenfeld & Nicolson, 1975).

9. Where a firm's investment expenditures are continually growing, provisions for accelerated depreciation can enable the firm to permanently reduce its tax payments rather than merely to shift them forward. In the limit 100% immediate capital write-off could enable the firm to avoid corporation taxes completely providing sufficient additional investment opportunities were available. See, for example, Johansen, *Public Economies* (1965) op. cit. pp. 251–6.

10. McDonald, 1967, op. cit. pp. 275–85.

11. The analysis of T. Page, *Conservation and Economic Efficiency*, Resources for the Future (Baltimore and London: Johns Hopkins University Press, 1977) pp. 112–13, is very similar to that presented below.

12. Our simplest model with perfect knowledge, profit-maximisation assumptions and a fixed stock of resource fairly predictably resulted in 'asset prices', 'royalties' or 'rents' being affected greatly by fiscal changes. Even in more realistic circumstances, however, there is evidence that such effects can be important. One interesting example concerns the response of state governments in Canada to suggested changes in the Federal Government's policy on 'expensing' provisions and depletion allowances. During 1971 and 1972 'provincial premiers vociferously opposed the proposed tax reforms' since they were expected to reduce the rents which provincial governments could expect to receive from the mineral companies. R. M. Hyndman and M. W. Bucovetsky further remark that 'the principal effect [of the Canadian tax system] was probably to enhance the rents paid to landowners, including provincial governments'. See 'Rents, Rentiers, and Royalties: Government Revenue from Canadian Oil and Gas', in E. W. Erickson and L. Waverman (eds), *The Energy Question: An International Failure of Policy* (University of Toronto Press, 1974) pp. 191–214.

13. This assumption concerning shifting is important. If the tax is shifted forwards and pre-tax rates of interest rise then, given similar treatment of capital gains, the increased market rate of interest will tend to advance depletion. Our provisional observations in Chapter 3, Section 3.5C, on taxation are therefore seen to apply to the case where the corporation tax is assumed to be shifted forwards.

14. Alfred Marshall used the famous example of a meteoric shower of exceptionally hard stones which would not wear out. He went on to consider the possibility of the stones being brittle, but only in the context of an inexhaustible store. See *Principles of Economics*, 8th ed. (London: Macmillan, 1925) book V, chap. ix.

15. The incidence of the property tax is a less settled issue than it seemed a few years ago. Its general regressivity in the context of the taxation of domestic property has been challenged recently by Henry Aaron, 'A New View of Property Tax Incidence', *American Economic Review* (May 1974) pp. 212–21. Following A. C. Harberger's analysis

of corporate income taxation ('The Incidence of the Corporation Income Tax', *Journal of Political Economy*, vol. 70, no. 3 (June 1962) pp. 215–40) he argues that the property tax has two effects on income distribution: the first via changes in the return on capital as capital moves from more to less highly taxed areas, and the second via the 'excise effect', i.e. induced changes in the prices of commodities. It seems that our depletion model produces an 'excise effect' in the direction of lower present prices and higher future ones.

16. A. G. Kemp, 'Fiscal Policy and the Profitability of North Sea Oil Exploitation', *Scottish Journal of Political Economy*, vol. 22, no. 3 (Nov 1975) pp. 244–60.

17. In the case of self-sufficiency Kemp argues that the burden would not be fully passed on to consumers but would be shared in traditional textbook fashion. This would appear to rely on the assumption that domestic producers are willing to sell in domestic markets at below the world price and that they cannot export any surplus at this world price. If exports are permitted and untaxed, a tax on domestic production and imports would result, as in the text, in no loss of rent, but in the self-sufficiency case there would be an increase in exports rather than a reduction in imports.

18. For a survey of these issues see Pechman, *Federal Tax Policy*, op. cit. pp. 135–41.

19. Kemp, op. cit., provides a more detailed appraisal of these taxes.

20. In the case of oilfields, for example, production rises gradually to a plateau and after some years at the plateau tapers off thereafter. With perfect capital markets the effects of tax measures on liquidity would be unimportant. Providing that a project had a positive net present value after tax, the precise timing of the tax payments would not alter the decision to develop the resource. In practice, however, legislators in the United Kingdom have taken care to avoid imposing a tax burden too early in a project's life, in case development should be discouraged (see Section 6.5).

21. Kemp, op. cit., provides some practical examples using data from four oilfields in the North Sea.

22. Kemp, op. cit. p. 251.

23. See D. I. MacKay and G. A. Mackay, *The Political Economy of North Sea Oil* (London: Martin Robertson, 1975) p. 42.

24. R. Garnaut and A. Clunies Ross, 'Uncertainty, Risk Aversion and the Taxing of Natural Resource Projects', *Economic Journal* (June 1975).

25. Garnaut and Clunies Ross include a numerical example of how the tax might be calculated in an appendix to their article.

26. M. T. Sumner, 'Progressive Taxation of Natural Resource Rents', *The Manchester School* (Mar 1978) pp. 1–16.

27. Sumner suggests a tax levied on net present value at a progressive rate and collected over a project's life. Mechanisms can be devised to

achieve this but they require knowledge of the firm's discount rate. As Sumner observes 'The obvious problem that arises is that the "firm's" discount rate is neither unique nor observable', op. cit. p. 12.

28. In the United Kingdom the rights of property-owners do not extend to the exploitation of all minerals which may be discovered, although this does not necessarily imply that owners will be unable to appropriate any of the rent which such a discovery might create.

29. This was established under the Convention on the Continental Shelf in 1958.

30. See B. D. Gardner, 'Towards a Disposal Policy for Federally Owned Oil Shales', in Gaffney (ed.), op. cit. pp. 169–95.

31. Studies have been carried out on offshore lease sales in the United States. Erickson and Spann, for example, have attempted to calculate expected bids for various tracts in the Gulf of Mexico. Their results indicate a close relationship between expected and actual bids. 'The bidding evidence suggests that the offshore leasing process is highly competitive.' See E. W. Erickson and R. M. Spann, 'The U.S. Petroleum Industry', in E. W. Erickson and L. Waverman (eds), *The Energy Question* (University of Toronto Press, 1974) pp. 5–24.

32. Especially in the United Kingdom (see Section 6.5, p. 173).

33. Economists with an instinctive regard for markets often tend to favour the competitive bidding system over more regulatory alternatives. K. W. Dam, for example, has consistently supported such a scheme in the North Sea while Gardner argues for a similar system in the development of oil shales. See K. W. Dam, 'Oil and Gas Licensing and the North Sea', *Journal of Law and Economics*, vol. 8 (Oct 1965) 51–75, and Gardner, 'Towards a Disposal Policy for Federally Owned Oil Shales', op. cit.

34. Extensive work has been undertaken in the United States on this aspect of mineral leasing policy. A useful survey of the major studies is given in F. M. Peterson and A. C. Fisher, 'The Exploitation of Extractive Resources: A Survey', *Economic Journal*, vol. 87, no. 4 (Dec 1977). pp. 681–721.

35. Quoted in F. M. Peterson, 'Two Externalities in Petroleum Exploration', in Brannon (ed.), *Studies in Energy Tax Policy*, op. cit. p. 102.

36. Erickson and Spann do not regard the existence of joint bidding as restricting competition. Rates of return on offshore operations they find to be competitive in the Gulf of Mexico. They also quote Professor J. W. Markham's evidence that joint bidding does not appear to reduce the number of bidders, op. cit.

37. The existence of risk-aversion is emphasised by Garnaut and Clunies Ross in their discussion of the advantages of the resource rent tax, op. cit. Henry Steele recommends a combination of lease bonus and severance tax for reasons of 'risk-sharing'. See 'Natural Resource Taxation: Resource Allocation and Distribution Implications'. So also does Gardner in the context of oil shales. Both in Gaffney (ed.), op. cit.

38. This is indeed a system which is common in North America.

The province of Alberta in Canada levies a royalty on natural-gas production as well as receiving lease bonuses at an auction.

39. This distinction between fiscal and non-fiscal instruments is perhaps rather arbitrary. Even some of the instruments termed 'non-fiscal' below may involve changing the flow of funds to the exchequer.

40. See, for example, P. E. Starratt and R. M. Spann, 'Dealing with the U.S. Natural Gas Shortage', in Erickson and Waverman (eds), *The Energy Question*, op. cit. pp. 25–46.

41. The model of Paul Davidson, Laurence Falk and Hoesung Lee suggests that changes in reserve–production ratios can be explained well by changes in interest rates and the well-head price on new contracts. They do not find a significant time trend for non-associated gas although the *overall* reserve production ratio does exhibit a negative time trend. They conclude that changes in reserve production ratios since 1955 'are primarily due to higher interest rates (or lower new well-head prices)'. See 'The Relations of Economic Rents and Price Incentives to Oil and Gas Supplies', in Brannon (ed.), op. cit. chap. 5, pp. 147–51. Starratt and Spann, op. cit. p. 34, report similar results from models by Spann and Erickson and MacAvoy and Pindyke: 'The supply of new natural gas discoveries is extremely sensitive to the well-head price of natural gas' (p. 34).

The influence of speculative behaviour should be borne in mind when considering these results. Davidson, Falk and Lee emphasise the 'user cost' of 'proving' reserves if de-regulation or higher Federal Power Commission prices are expected.

42. K. W. Dam, The Evolution of North Sea Licensing Policy in Britain and Norway', *Journal of Law and Economics*, vol. 17, no. 2 (Oct 1974), esp. pp. 221–6, and vol. 13, no. 1 (Apr 1970) pp. 11–44. For a detailed account of the negotiations between the Gas Council and the Oil Companies during the 1960s see 'The Pricing of North Sea Gas in Britain', *Journal of Law and Economics* (1970), by the same author. See also M. B. Posner, *Fuel Policy* (London: Macmillan, 1973) chap. 11, for a discussion of the determination of the U.K. price of natural gas.

43. The general functions of the B.N.O.C. are laid down in section 2 of the Petroleum and Submarine Pipelines Act 1975. The reasons for setting up a fully fledged oil company probably lie more in the realm of politics than economics and would take us beyond the scope of this chapter. Dam, 1974, op. cit., discusses in more detail the reasons for the setting-up of Statoil.

44. Dam, 1974, op. cit. p. 259.

45. MacKay and Mackay discuss the possibilities of an 'appreciation' in the estimated reserves. They argue that reserves will be uprated over time 'probably by not less than 25% on average', op. cit. p. 62. See Chapter 2 above and Table 2.10.

46. In the fourth round of licensing £21,050,001 was paid for block 211/21 with no results to date.

47. Pipelines have been laid, for example, from the Forties field to

Cruden Bay and from Ninian and Brent to Sullom Voe in Shetland. Gas from the Frigg field is landed at St Fergus.

48. A breakdown of estimated capital and operating costs of the various fields in the North Sea is given in A. G. Kemp, 'The Taxation of North Sea Oil', University of Aberdeen, *North Sea Study Occasional Papers* (Aug 1976) appx I.

49. Points (*a*), (*b*) and (*c*) from *Energy Policy: A Consultative Document*, Cmnd 7101 (London: H.M.S.O., 1978) pp. 35–6.

50. *UK Offshore Oil and Gas Policy*, Cmnd 5696 (London: H.M.S.O., July 1974).

51. *Serving the Offshore Industry*, Dept of Energy (London: H.M.S.O., 1976).

52. We are here referring to production licences which give the licensee the right to extract any resource he discovers. There also exist quite separate exploration licences. Details of sixth-round payments are given in *UK Offshore Petroleum Production Licensing, Sixth round, A Consultative Document*, Dept of Energy, appx 2.

53. These conditions apply to the sixth round taking place in 1978. However, apart from B.N.O.C. participation which began in the fifth round, few changes have occurred.

54. Sixth Round Consultative Document, op. cit., appx 3.

55. Petroleum and Submarine Pipelines Act (London: H.M.S.O., section 41 (3)).

56. The Oil Taxation Act 1975 (London: H.M.S.O., 1975).

57. Two important studies of the effects of the oil taxation system are Kemp, 'The Taxation of North Sea Oil', op. cit., and J. R. Morgan and C. Robinson, 'The Comparative Effects of the U.K. and the Norwegian Oil Taxation System on Profitability and Government Revenue', *Accounting and Business Research* (Winter 1976). An abridged version of the latter can be found as 'A Comparison of Tax Systems', *The Petroleum Economist* (May 1976).

58. Kemp estimates that the oil allowance raises the internal rate of return to Auk and Argyll by about 10 percentage points (from 39% to 48% in the case of Auk). The Brent field, in contrast, is unchanged with an internal rate of return of 19% (op. cit. table VI, p. 19).

59. Kemp, 1976, op. cit. p. 18.

60. Statement made by the Minister of Mineral Resources to the Legislative Assembly of Saskatchewan, 14 Apr 1976.

61. See Uranium Royalty Regulations, Order-in-Council 1090/76, 27 July 1976..

62. The tax follows the 'slice' principle not the 'slab'. For example, a mine with investment of £1 million and operating profit of $200,000 would pay no tax on £150,000 plus 15% on the next $50,000 = £7500.

63. I.e. 'Chartered Bank Lending Rates on Prime Business Loans', published in the *Bank of Canada Review*.

64. The interested reader is referred to the following surveys of U.S. tax policy: 'Existing Tax Differentials and Subsidies Relating to the

Energy Industries, Brannon in *Studies in Energy Tax Policy*, op. cit. pp. 3–40 (1975); 'Tax Incentives in the U.S. Petroleum Industry', Millsaps, Spann and Erickson in *The Energy Question*, op. cit. pp. 99–122, and Page, *Conservation and Economic Efficiency*, Resources for the Future, op. cit. (1977) chap. 6.

65. By 1984 it should be 15 per cent on the first 1000 barrels of oil per day.

66. For an interesting and fuller account of the history of the depletion allowance see Page, op. cit. chap. 6. W. Vickrey, 'Economic Criteria for Optimum Rates of Depletion', in Gaffney (ed.), op. cit. pp. 315–30, also contains a historical account of the special tax provisions accorded to the mineral industries.

67. Pechman, *Federal Tax Policy*, 3rd ed., op. cit. p. 152.

68. Page, op. cit. pp. 122–4. Another important tax provision causing much controversy in the United States is the Foreign Tax Credit. This is simply a provision which enables companies to credit foreign taxes against U.S. taxes. If income taxes are levied by foreign governments at greater rates than those applying in the United States, the company will not have to pay U.S. taxes on the income concerned. In the case of petroleum, however, many of the taxes paid to foreign governments are not strictly income taxes but are more in the nature of 'royalties' or 'rents' for working the deposits. It is argued therefore that these payments are costs of production, equivalent to lease payments, and should be treated like other costs, i.e. deducted from gross income in computing tax liability, but not fully credited against domestic tax payments. For a detailed study of the foreign operations of U.S. petroleum companies see Glenn P. Jenkins, 'United States Taxation and the Incentive to Develop Foreign Primary Energy Resources', in Brannon (ed.), op. cit. pp. 203–45.

69. A depletion allowance of S1 can now be claimed for each $3 of exploration and development expenditure up to a limit of $33\frac{1}{3}\%$ of taxable profit, i.e. depletion must be 'earned'.

70. See A. Harberger, 'The Taxation of Mineral Industries', *Federal Tax Policy for Economic Growth and Stability*, 84 Cong., 1 Sess., Nov 1955.

71. See S. L. McDonald, 'Percentage Depletion and the Allocation of Resources: The Case of Oil and Gas', *National Tax Journal* (Dec 1961) pp. 323–36. His analysis produced a stream of subsequent papers in *National Tax Journal*: D. H. Eldridge, 'Rate of Return, Resource Allocation and Percentage Depletion' (June 1962) pp. 209–17; R. A. Musgrave, 'Another Look at Depletion' (June 1962) pp. 205–8; and P. O. Steiner, 'The Non-Neutrality of Corporate Income Taxation – With and Without Depletion' (Sep 1963) pp. 238–51.

72. McDonald provides a résumé of the case against him in 'Percentage Depletion, Expencing of Intangibles, and Petroleum Conservation', in Gaffney (ed.), op. cit. pp. 280–5.

73. Stiglitz, 'The Efficiency of Market Prices in Long-run Allocations in the Oil Industry', in Brannon (ed.), 1975, op. cit. pp. 73–8.

74. See, for example, Musgrave and Musgrave, 1976, op. cit. pp. 494–6.

75. A good general survey is provided by H. W. Richardson, *Economic Aspects of the Energy Crisis* (Farnborough, Hants: Lexington Books/Saxon House, 1975) pp. 64–9, pp. 132–6.

76. For example, J. C. Cox and W. W. Wright, 'The Cost-effectiveness of Federal Tax Subsidies for Petroleum Reserves: Some Empirical Results and their Implications', in Brannon (ed.), op. cit. pp. 177–202; Millsaps, Spann and Erickson discuss the results of their model in 'Tax Incentives in the U.S. Petroleum Industry', op. cit. They prefer the strategic stockpile as a solution to the problem of security.

77. S. L. McDonald, 'US Depletion Policy: Some Changes and Likely Effects', *Energy Policy* (Mar 1976) pp. 56–62.

78. In the Petroleum and Submarine Pipelines Act 1975. See Chapter 3 for a further discussion of U.K. depletion policy.

## Chapter 7

1. Most modern textbooks cover this area well, e.g. W. Nicholson, *Micro Economic Theory* (New York: Dryden Press, 1978) chap. 6, pp. 147–78.

2. The von Neumann and Morgenstern axioms are effectively an addition to the standard axioms of consumer theory to take account of uncertainty.

(*a*) For any two alternatives (say $C_1$, $C_2$) either $C_1 P C_2$ or $C_2 P C_1$ or $C_1 I C_1$, where $P$ denotes 'is preferred to' and $I$ denotes indifference.

(*b*) All pairs of alternatives having been ranked according to (*a*) the final complete ordering must be transitive, i.e. if $C_1 P C_2$ and $C_2 P C_3$, then it must follow that $C_1 P C_3$. If the consumer ranks $C_3 P C_1$ the transitivity axiom is denied.

(*c*) If $C_1 P C_2 P C_3$ there will exist some probability $r$, where $0 < r < 1$, such that the individual is indifferent between $C_2$ with certainty and a lottery ticket offering $C_1$ or $C_3$ with probabilities $(1 - r)$ and $r$ respectively.

(*d*) If $C_1 I C_2$ and if the individual is offered a choice of lottery tickets, one offering $C_1$ or $C_3$ with probabilities $(1 - r)$ and $r$ respectively, and the other offering $C_2$ or $C_3$ with the same respective probabilities, the individual will be indifferent between the two tickets.

(*e*) If $C_1 P C_2$ and the individual is offered two lottery tickets involving these two outcomes, the individual will chose the ticket with the highest probability of obtaining $C_1$.

(*f*) The 'axiom of complexity' is an extension of (*e*) and asserts broadly that an individual will be indifferent between lottery tickets offering $C_1$ or $C_2$ with given probabilities *irrespective of the complexity of the gamble*. As an example, a person would be indifferent between a simple gamble offering £12 or £4 with probabilities $\frac{1}{4}$ and $\frac{3}{4}$ and a complex gamble in which the individual has a 50—50 chance of choosing a lottery ticket offering £12 or £4 with probabilities 0 and 1 or a ticket offering £12 or £4 with probabilities $\frac{1}{2}$ and $\frac{1}{2}$. In either case the probability distribution of final outcomes will be the same, £12 with probability $\frac{1}{4}$ and £4 with probability $\frac{3}{4}$.

For a fuller discussion of these axioms and examples of their use (especially for axiom *c*) in constructing a utility function see H. A. John Green, *Consumer Theory*, rev. ed. (London: Macmillan, 1976) chap. 13, and technical appendix T.9. The classic reference in this field is J. von Neumann and O. Morgenstern, *The Theory of Games and Economic Behaviour* (Princeton, N.J.: Princeton University Press, 1944). See also A. A. Alchian, 'The Meaning of Utility Measurement', *American Economic Review*, vol. 43, no. 1 (1953), reprinted in W. Breit and H. M. Hochman (eds), *Readings in Microeconomics* (New York: Holt, Rinehart & Winston (1967) pp. 69—88.

3. For a more detailed exposition see J. Hirshleifer, 'Investment Decisions under Uncertainty: Choice-Theoretic Approaches', *Quarterly Journal of Economics*, vol. 79, no. 4 (1965) pp. 509—36.

4. This is easily seen by inspecting expression (7.4). Minus the left-hand side of the expression, represents the slope of the indifference curve. Letting $\bar{C}_{1a} = \bar{C}_{1b}$, $V'(\bar{C}_{1b}) = V'(\bar{C}_{1a})$ and we are left with the ratio of state probabilities.

5. The present analysis follows closely that of J. Hirshleifer and D. L. Shapiro, 'The Treatment of Risk and Uncertainty', *Quarterly Journal of Economics*, vol. 77, no. 4 (1963) pp. 505—30.

6. Ibid.

7. Fear of the takeover raid may result in managers wishing to avoid low returns. See, e.g., R. Marris, *The Economic Theory of 'Managerial' Capitalism* (London: Macmillan, 1964).

8. The problem is simply to find $\lambda$ such that:

$$\frac{1}{2}\sqrt{(144 - \lambda)} + \frac{1}{2}\sqrt{81} = 10$$
$$\sqrt{(144 - \lambda)} = 11$$
$$144 - \lambda = 121$$
$$\lambda = 23$$

9. For each company the outcomes will be £98.1 m. or £104.4 m. In utility terms these will be $U_1 = 9.9$ and $U_2 = 10.22$.
Thus $E(W) = 101.25$ and $E(U) = 10.06$.
The certainty-equivalent will be $[E(U)]^2 = 101.225$. Hence each firm's cost of risk will be $101.25 - 101.225 = 0.025$.
Total cost of risk-bearing is thus 0.25.

10. For a definition of Pareto optimality under condition of uncertainty and the derivation of appropriate rules for public investments see

A. Sandmo, 'Discount Rates for Public Investment under Uncertainty', *International Economic Review*, vol. 13, no. 2 (June 1972) pp. 287–302.

11. Op. cit. p. 295.

12. For a statement of this view see the discussion of a paper by Harberger by P. A. Samuelson and W. Vickrey, *American Economic Review, Papers and Proceedings* (1964) pp. 88–96. The 'law of large numbers' can be stated in various ways. Let $R_i$ represent the return to a particular project. Suppose there are $n$ projects and that $R_1, R_2, \ldots R_n$ are identically distributed independent random variables with finite mean and variance. Further let $E(R_i) = \mu$, $V(R_i) = 6^2$ and define $\bar{R} = 1/n(R_1 + R_2 + \cdots R_n)$. By increasing the number of projects undertaken it can be shown that $\bar{R}$ the average return to each project 'converges' to $\mu$. More formally:

$$P[\ |\ \bar{R} - \mu\ | < e] \geqslant 1 - \frac{6^2}{e^2 n},$$

where $e$ is any positive number. Clearly $e$ can be assumed arbitrarily small but a sufficiently large $n$ will cause the right-hand side to approach unity. Thus we can, weith large enough $n$ be 'virtually certain' that the average project return will be 'close' to $\mu$ the expected return to each project.

13. K. J. Arrow and R. C. Lind, 'Uncertainty and the Evaluation of Public Investment Decisions', *American Economic Review*, vol. 60. (June 1970) pp. 364–78.

14. Thus: 'When the government undertakes an investment, each taxpayer has a small share of that investment with the returns being paid through changes in the level of taxes.' Arrow and Lind, op. cit.

15. The point here is that there may be no finite sum which would compensate a person for severe bodily mutilation or even death. If this is the case, no amount of redistribution via the fiscal system will succeed in spreading risk.

16. We might write this more formally as:

$$P(X_i/Y_j) = P(X_i) \text{ for all } i, j.$$

17. This, of course, ignores the problem of 'public bads' discussed earlier which will be important in the case of nuclear energy. There may also be significant 'irreversibilities' attached to decisions to produce nuclear energy and these will be discussed in Section 7.5, p. 216.

18. Sandmo, 1972, op. cit. p. 300.

19. For a more detailed discussion of this case see. A. C. Fisher and J. V. Krutilla, 'Valuing Long-Run Ecological Consequences and Irreversibilities', in H. Peskin and E. Seskin (eds), *Cost Benefit Analysis and Water Pollution Policy* (Washington, D.C.: The Urban Institute, 1975) pp. 271–90.

20. C. Henry, 'Irreversible Decisions under Uncertainty', *American Economic Review*, vol. 64, no. 6 (Dec 1974) pp. 1006–12.

21. Actinides comprise elements following actinium in the periodic table. They include uranium, neptunium, plutonium and curium.

22. For more details see Final Report by the Energy Policy Project of the Ford Foundation, *A Time to Choose* (New York: Ballinger, 1974) pp. 208–10. Also Royal Commission on Environmental Pollution, 6th Report, *Nuclear Power and the Environment*, Cmnd 6618 (London: H.M.S.O., 1976) chap. 8.

23. B. A. Weisbrod, 'Collective-Consumption Services of Individual Consumption Goods', *Quarterly Journal of Economics*, vol. 78 (Aug 1964) pp. 471–7.

24. C. J. Cicchetti and A. Myrick Freeman III, 'Option Demand and Consumer Surplus: Further Comment', *Quarterly Journal of Economics*, vol. 85 (Aug 1971) pp. 528–39.

25. K. J. Arrow and A. C. Fisher, 'Environmental Preservation, Uncertainty and Irreversibility', *Quarterly Journal of Economics*, vol. 88 (May 1974) pp. 312–19. The exposition which follows relies heavily on this article and the work of Fisher and Krutilla, op. cit.

26. The original article by Arrow and Fisher allows for this possibility.

27. A precisely equivalent process of reasoning establishes the same result in the case for which $E[B_{d2} - C_{d2}] > E[B_{p2}]$.

28. The 'risk-margin' refers to the rate at which individuals are prepared to trade uncertain for certain claims. We noted in expression (7.8) that in a two-period analysis the price of a certain dollar next period

$$P_1 = P_{1a} + P_{1b} = \frac{1}{1+r},$$

where $a$ and $b$ were two possible states of the world and $r$ was the riskless rate of interest.

Consider an investment project with present certainty-equivalent value $W^*$:

$$W^* = -K_0 + P_{1a}R_{1a} + P_{1b}R_{1b}.$$

Given that individuals aim to maximise the certainty equivalent of their wealth holdings, *marginal* projects will have a certainty equivalent present value of zero, i.e.

$$K_0 = P_{1a}R_{1a} + P_{1b}R_{1b}.$$

Suppose now that the investment under consideration is a class for which returns are zero if state of the world $b$ occurs and positive if state of the world $a$ occurs. We would then have, at the margin,

$$K_0 = P_{1a}R_{1a}$$

$$1 = P_{1a}\frac{R_{1a}}{K_0} = P_{1a}R_{1a}^*,$$

where $R^*_{1a}$ = Returns in state of the world $a$ per dollar of investment. The *expected* gross return per dollar of investment will be $\Pi_{1a}R^*_{1a}$. From expression (7.9) we then find the 'risky rate of interest' for investments in this class as

$$1 + p = \frac{\Pi_{1a}R^*_{1a}}{P_{1a}R^*_{1a}} = \Pi_{1a}R^*_{1a}.$$

However, it is already known that since, in this simple case, returns occur only in state of the world $a$, the 'risky rate of interest' is also defined by the relationship

$$1 + p = \frac{\Pi_{1a}}{P_{1a}}.$$

But $\qquad \dfrac{1}{P_{1a}} = \dfrac{1}{P_1} \cdot \dfrac{P_1}{P_{1a}} = (1 + r)\dfrac{P_1}{P_{1a}}.$

Thus equilibrium will be characterised by the following equality:

$$R^*_{1a} = (1 + r)\frac{P_1}{P_{1a}}.$$

The term $P_1/P_{1a}$ represents the 'risk margin' for this class of investment.

Given the existence of perfect markets, all consumers will equate their marginal rates of substitution between uncertain claims and sure claims to future consumption to this price ratio. (This result is easily derived from the first-order conditions leading to equation (7.4) in the text.) Thus all consumers will be willing to sacrifice the same number of state $a$ contingent claims to \$1 next period for a single claim to a *certain* dollar. Risk margins will be the same for all individuals.

29. See Sandmo, op. cit. (1972). Sandmo demonstrates that in an economy offering no possibility of portfolio diversification and hence risk-pooling, the risk margin on public-sector projects in risk class $k$ should be 'a weighted harmonic mean of all consumers' risk margins for investment of risk class $k$, the weights being consumers' shares of the surplus in the public sector'. (p. 299).

30. For a description of 'Safety Analysis' see the Royal Commission on Environmental Pollution, 6th Report, op. cit. pp. 108–11. The basic idea is that all possible ways in which accidents can occur are identified and the probability of each is calculated from the reliability of the various mechanical components involved. Critics point to the fact that there is no way of knowing that *all* possibilities have been considered.

31. Hirshleifer, 1965, op. cit. p. 253.

32. For excellent and fuller surveys in this field see W. J. Baumol, *Economic Theory and Operations Analysis*, 2nd ed. (Eaglewood Cliffs, N.J.: Prentice-Hall, 1965) chap. 24. Also E. J. Mishan, *Cost Benefit*

*Analysis* (London: Allen & Unwin, 1971); and R. Dorfman, 'Decision Rules under Uncertainty', in *Cost-Benefit Analysis*, ed. R. Layard (Harmondsworth: Penguin Books, 1972), reprinted from 'Basic Economic and Technologic Concepts: a general statement', in A. Maass *et al.*, *Design of Water Resource Systems* (Cambridge, Mass.: Harvard University Press, 1962) pp. 129–58.

33. See *Committee on Principles of Decision Making for Regulating Chemicals in the Environment* (Washington, D.C.: National Academy of Sciences, 1975) appx H, p. 182.

34. See ibid. appx G.

35. Experience suggests that the behaviour of prototypes and commercial versions are often very different. The advanced gas-cooled reactor (AGR) in the United Kingdom was developed from a prototype built at Windscale of 33 MW. The commercial stations produced 660 MW and enormous design and construction problems were encountered. As David Fishlock has written: 'their most ardent supporters agree that the leap from 33 to 660 MW was too big and that the first pair of AGR stations, Hinkley B and Hunterston B have really been demonstration projects – at around £140 m apiece'. See D. Fishlock, 'Pitfalls for the Energy Planners', *Financial Times*, 25 Jan 1978.

36. Op. cit. p. 183.

37. See Chapter 5, Section 4.

38. See Royal Commission on Environmental Pollution, 6th Report, op. cit. Chap. VI. The 'curie' is a unit of measurement of radioactivity. One curie corresponds to the activity displayed by one gram of radium $-226$, i.e. $3.7 \times 10^{10}$ disintegrations per second.

39. See *Energy Policy: A Consultative Document*, Cmnd 7101 (London: H.M.S.O., Feb 1978). Unfortunately, it is not at all clear which decision rule they had in mind. As was noted in Section 7.7 the minimax regret strategy does not imply returns which are relatively stable or robust over states of the world. The Green Paper seems to imply that the whole distribution of returns over all states is important, but if this is so the term 'minimum regret' has no obvious meaning. In terms of Section 7.7 it would appear that strategy 3 is most 'robust' in the Green Paper's terms since it produces zero regret for two out of three states of the world. What would happen if there existed no strategy having such an advantage over the others is unclear.

# Chapter 8

1. See, for example, the various papers in J. A. G. Thomas (ed.) *Energy Analysis* (Guildford: IPC Science and Technology Press, 1978).

2. H. T. Odum, *Ambio*, vol. 2 (1973) p. 220.

3. The quotation is taken from D. A. Huettner, 'Net Energy Analysis: An Economic Assessment', *Science*, vol. 192 (9 Apr 1976) p. 101.

4. See, for example, M. W. Gilliland, 'Energy Analysis and Public Policy', *Science*, vol. 189 (26 Sep 1975) pp. 1051–6.

5. P. F. Chapman, 'Conventions, Methods and Implications of Energy Analysis', a paper presented to the Institute of Fuel and the Operational Research Society Joint Conference 'Understanding Energy Systems', session C, *Energy Accounting* (London: Apr 1975).

6. Ibid.

7. See, for example, D. J. Wright, 'The Natural Resource Requirements of Commodities', *Applied Economics*, vol. 7 (1975) pp. 31—9.

8. Chapman, 'Conventions, Methods and Implications of Energy Analysis', op. cit. p. 38.

9. Enthalpy refers to available energy and is distinguished from entropy, which is unavailable energy. The first law of thermodynamics tells us that energy can be neither created nor destroyed. The second law tells us that the entropy of a closed system increases continuously towards a maximum.

10. P. F. Chapman, 'Energy Costs: A Review of Methods', *Energy Policy*, vol. 2, no. 2 (June 1974) p. 93.

11. G. Leach, 'Net Energy Analysis — Is It Any Use?', *Energy Policy*, vol. 3, no. 4 (Dec 1975) pp. 340—1.

12. Leach, ibid. p. 338, has argued that the arbitrary choice of the sub-system will make Public Law 03—577 in the United States virtually unworkable since every published study will be open to reasonable objections.

13. This suggestion was made, for example, by a number of the working groups at the Net Energy Analysis Workshop, Institute of Energy Studies, Stanford University, California, 25—28 Aug 1975. See *Draft Proceedings Report*, Reports of Working Groups 1—A, 1—B, and 1—C, Quoted in Leach, 'Net Energy Analysis', op. cit.

14. The calorific value is a measure of the heat energy which is potentially available from a fuel. This measures the maximum energy which can be extracted from a fuel.

15. C. D. Kylstra, 'Energy Analysis as a Common Basis for Optimally Combining Man's Activities and Nature's', chap. 17 in G. F. Rohrlich (ed.), *Environmental Management* (New York: Ballinger, 1976) p. 274.

16. See M. G. Webb and David Pearce, 'The Economics of Energy Analysis', *Energy Policy*, vol. 3, no. 4 (Dec 1975).

17. Chapman, 'Energy Costs: A Review of Methods', op. cit. p. 91.

18. Chapman, 'Energy Analysis: A Review of Methods and Applications', *Omega*, vol. 4, no. 1 (Feb 1976) pp. 19—33.

19. Gilliland, op. cit. p. 1053. On p. 1056 of this article the author writes: 'Dollar evaluations also change with time due to the changing value of money and assumptions concerning, for example, the discount rate. For a specific technology, such as the present nuclear fuel cycle and its supporting techniques, the energy evaluation will not change with time.'

20. See Webb and Pearce, op. cit. p. 324.

21. See R. Turvey and A. R. Nobay, 'On Measuring Energy Consumption', *Economic Journal* (Dec 1965).

22. See, for example, P. Chapman and N. D. Mortimer, *Energy Inputs and Outputs of Nuclear Power Stations*, The Open University Energy Research Group, Report ERG 005 (1974).

23. B. Hannon, 'An Energy Standard of Value', *The Annals of the American Academy of Political and Social Science*, vol. 405 (1973).

24. See, for example, Hannon, op. cit. p. 140, and Gilliland, op. cit. Some authors are less explicit, but they imply that in the choice of both output mixes and input combinations consideration should be given to energy efficiency. See, for example, M. Slesser, 'Energy Analysis in Policy Making', *New Scientist* (1973) pp. 328–9; and G. Leach, 'The Energy Costs of Food Production', in A. Bourne (ed.), *The Man– Food Equation* (London: Academic Press, 1976).

25. The following analysis follows that of D. A. Huettner, 'Net Energy Analysis: An Economic Assessment', *Science*, vol. 192 (9 Apr 1976) pp. 101–4.

26. Hannon, op. cit.

27. Huettner, op. cit. p. 103.

28. G. W. Edwards, 'Energy Budgeting: Joules or Dollars', *Australian Journal of Agricultural Economics* (Dec 1976) p. 186.

29. See Webb and Pearce, op. cit. pp. 326–30.

30. P. F. Chapman, 'The Relation of Energy Analysis to Cost Analysis,' paper presented to the Institution of Chemical Engineers working party on materials and energy resources (1975).

31. On this general question of growth-rate scenarios see P. F. Chapman, *Fuel's Paradise* (Harmondsworth: Penguin Books, 1975).

32. On the problems relating to the choice of the optimal terminal capital stock see J. de V. Graaff, *Theoretical Welfare Economics* (London: Cambridge University Press, 1957) chap. 6.

33. In economics it is usual to stress the time-dimension when defining a commodity. This means that 1 therm of natural gas in 1980 is not considered to be identical to 1 therm in 1990 or 2000. Thus to make these therms comparable it is necessary to express them as if they were all available at a common date.

34. For example, Gilliland, op. cit. p. 1056.

35. For some examples see Chapman, *Fuel's Paradise*, op. cit. pp. 118–19.

36. Hannon, 'An Energy Standard of Value', *The Annals of the American Academy of Political and Social Science*, vol. 405 (1973).

37. See, for example, Chapman and Mortimer, *Energy Inputs and Outputs of Nuclear Power Stations*, The Open University Energy Research Group Report ERG 005 (1974). For a critique of this and similar studies see L. G. Brookes, 'Energy Accounting and Nuclear Power', *Atom*, no. 227 (London: U.K. Atomic Energy Authority, Sep 1975).

38. See. E. J. Mishan, 'A Normalisation Procedure for Public Investment Criteria', *Economic Journal* (1967).

## Chapter 9

1. The best-known studies are J. Forester, *World Dynamics* (Cambridge, Mass.: Wright-Allen Press, 1971), and D. H. Meadows *et al.*, *The Limits to Growth* (London: Earth Island, 1972).

2. See especially W. D. Nordhaus, 'World Dynamics: Measurement without Data', *Economic Journal* (Dec 1973). Also J. A. Kay and J. A. Mirrlees, 'The Desirability of Natural Resource Depletion', in D. W. Pearce (ed.), *The Economics of Natural Resource Depletion* (London: Macmillan, 1975), and H. W. Richardson, *Economic Aspects of the Energy Crisis* (Farnborough, Hants: Lexington Books, Saxon House, 1975) chap. 2.

3. On the importance of the choice of the appropriate administrative and organisational structures for the determination and implementation of energy policy see P. L. Cook and A. J. Surrey, *Energy Policy* (London: Martin Robertson, 1978) chap. 12.

4. *The National Plan*, Cmnd 2764 (London: H.M.S.O., Sep 1965) chap. 11.

5. *Fuel Policy*, Cmnd 2798 (London: H.M.S.O., Oct 1965 para. 16. In 1956 power stations used less than 0.7 million tons of coal equivalent oil, but by 1959 as a result of the programme of conversion to oil this had increased to 7.2 million tons.

6. In 1955 the United Kingdom imported 11½ million tons of coal, of which nearly 50 per cent came from the United States.

7. It was still imposed in 1978 at the level of approximately 1.0 p a gallon.

8. *Fuel Policy*, Cmnd 2798, op. cit.

9. Ibid. para. 1.

10. Ibid. paras 18, 51 and 82.

11. *Fuel Policy*, Cmnd 3438 (London: H.M.S.O., Nov 1967) para. 4.

12. See C. I. K. Forster, *The Statistical Basis to National Fuel Policy*, a paper presented to the Institute of Actuaries, London, on 27 January 1969.

13. Ibid. p. 5.

14. *Fuel Policy*, Cmnd 3438, op. cit. para. 15.

15. See M. G. Webb, 'Some Aspects of Nuclear Power Economics in the United Kingdom', *Scottish Journal of Political Economy*, vol XV (Feb 1968).

16. For a detailed critique of this AGR programme see D. Burn, *Nuclear Power and the Energy Crisis* (London: Macmillan, 1978).

17. *Fuel Policy*, Cmnd 3438, op. cit. para. 47.

18. *Energy Policy: A Consultative Document*, Cmnd 7101 (London: H.M.S.O., 1978).

19. By 'adequate' was meant that energy supplies should not be a constraint on economic growth.

20. *Energy Policy: A Consultative Document*, op. cit. para. 2.9, p. 4.

21. Ibid. para. 2.11, p. 5; see also para. 15.5 (vi), p. 82 and appx 4.

22. Ibid. para. 14.18, p. 75.

23. Ibid. para. 4.20, p. 75.

24. Ibid. p. 77.

25. For a discussion of British offshore oil policy see 'Exploration and Development of U.K. Continental Shelf Oil', *Energy Commission Paper No. 17* (London: Department of Energy, Oct 1978). See also the discussion in Chapter 6 (pp. 169–79) above.

26. In 1978 nineteen member countries of the O.E.C.D. belonged to the I.E.A., but its membership is not conterminous with that of the O.E.C.D. For example, neither France nor Australia are members.

27. A number of member countries reserved their position with regard to the promotion of nuclear-generating capacity. These countries were Denmark, the Netherlands, New Zealand, Norway, Spain and Sweden.

28. *The National Energy Plan* (Washington: Executive Office of the President, Apr 1977).

29. There is a large literature relating to the economic consequences of the regulation by the Federal Government of natural-gas prices. The interested reader is recommended to look at P. W. Macavoy and R. S. Pindyck, *The Economics of the Natural Gas Shortage (1960–1980)* (Amsterdam and New York: North-Holland/American Elsevier, 1975); P. W. Macavoy and R. S. Pindyck, 'Price Controls and the Natural Gas Shortage', in *Perspectives on U.S. Energy Policy: A Critique of Regulation*, ed. E. J. Mitchell (New York: Praeger 1976); and Milton Russell, 'Natural Gas Curtailments: Administrative Rationing or Market Allocation', in *New Dimensions in Public Utility Pricing*, ed. H. M. Trebing (East Lansing, Mich.: Michigan State University Press, 1976).

30. See R. E. Hall and R. S. Pindyck. 'The Conflicting Goals of National Energy Policy', *The Public Interest*, no. 47 (Spring 1977).

31. It is worth noting that with U.S. oil prices being lower than world prices this policy should have given U.S. industry an export advantage.

32. *The National Energy Plan*, op. cit. p. xiii.

33. Ibid. p. xi.

34. Ibid. p. xvii.

35. Ibid. p. 46.

36. *A National Energy Plan for the United States*, an address to a Joint Session of Congress by President Carter on 20 Apr 1977, p. 8.

37. See Walter Goldstein, 'The Politics of U.S. Energy Policy', *Energy Policy*, vol. 6 (Sep 1978) no. 3, pp. 180–95.

38. See *The Economist*, 21 Oct 1978, p. 36.

39. Beginning with the 1980 models cars doing 14–15 mpg will be taxed $200 and those doing less than 13 mpg $550. Both the mpg threshold and the tax will rise annually.

40. On this paradox see, for example, R. G. Lipsey, *An Introduction to Positive Economics*, 4th ed. (London: Weidenfeld & Nicolson, 1975) pp. 168–70.

# INDEX